THE MAGNETIC UNIVERSE

J. B. ZIRKER

THE
MAGNETIC

THE ELUSIVE TRACES OF AN INVISIBLE FORCE

UNIVERSE

THE JOHNS HOPKINS UNIVERSITY PRESS

BALTIMORE

© 2009 The Johns Hopkins University Press
All rights reserved. Published 2009
Printed in the United States of America on acid-free paper

2 4 6 8 9 7 5 3 1

The Johns Hopkins University Press
2715 North Charles Street
Baltimore, Maryland 21218-4363
www.press.jhu.edu

Library of Congress Cataloging-in-Publication Data
Zirker, Jack B.
The magnetic universe : the elusive traces of an invisible force / J.B. Zirker.
p. cm.
Includes bibliographical references and index.
ISBN-13: 978-0-8018-9301-8 (hardcover : alk. paper)
ISBN-10: 0-8018-9301-1 (hardcover : alk. paper)
ISBN-13: 978-0-8018-9302-5 (pbk. : alk. paper)
ISBN-10: 0-8018-9302-X (pbk. : alk. paper)
1. Magnetic fields.   2. Cosmic magnetic fields.   3. Magnetism.
4. Magnetosphere.   5. Heliosphere (Ionosphere)   6. Gravity.   I. Title.
QC754.2.M3Z57 2009
538—dc22   2008054593

A catalog record for this book is available from the British Library.

The last printed pages of the book are an extension of this copyright page.

Special discounts are available for bulk purchases of this book. For more information,
please contact Special Sales at 410-516-6936 or specialsales@press.jhu.edu.

The Johns Hopkins University Press uses environmentally friendly book materials,
including recycled text paper that is composed of at least 30 percent post-
consumer waste, whenever possible. All of our book papers are acid-free, and
our jackets and covers are printed on paper with recycled content.

CONTENTS

PREFACE

Imagine that Federal Express delivers a package to you. There is no return address on the package, so you open it warily. Inside, you find a pair of dark eyeglasses. You put them on, and immediately you are in total darkness. But as your eyes adapt, you see that you are immersed in a thicket of shimmering threads. Everywhere you turn you see them. As you turn and move about the room, they part to let you pass, but you feel nothing. You notice that the bulb in your floor lamp is sheathed in flickering coils of light. When you touch the bulb gingerly, all the threads in the room vibrate.

Now you go outdoors and look up at the night sky. The whole dome of the sky is covered with a veil of parallel shimmering threads that line up north to south. When you walk back indoors, an idea occurs to you. You pull out your key chain, from which hangs a toy compass. Sure enough, the north and south ends of the compass needle are linked by faint loops.

Eureka! Now you understand what's happening. These strange eyeglasses allow you to see magnetic fields. They are everywhere, unnoticed but detectable with proper equipment.

Indeed, the entire universe is permeated with magnetic fields. We can't see them, because unfortunately no magical eyeglasses exist. But astrophysicists can detect them by the effects they have on visible light, radio waves, and x-rays. In stars like our Sun, the fields are thousands of times stronger than Earth's. In pulsars and white dwarfs, the fields are millions of times stronger. Between the arms of our Milky Way Galaxy they are millions of times weaker. The weakest fields of all connect galaxies within a cluster and stretch across immense voids to reach distant clusters.

Magnetic fields work in many wondrous ways during the evolution of the universe. They store and release energy, sometimes catastrophically, as in solar and stellar flares. They channel flows of ionized gases, as in pulsar jets and stellar winds. They assist in forming stars and planetary systems from the interstellar gas. They may even help to form galaxies. And incidentally, they keep us Earthlings safe from the bombardment of cosmic rays. One could argue that magnetism is second in importance only to gravity as an agent of change in the universe.

Just to whet your appetite, take a look ahead at figure 10.4. It shows a double radio source (Cygnus A) that consists of two gigantic lobes connected by a faint thread to a tiny star-like object at the center. This dumbbell-shaped object is immense; the lobes are separated by about 10 million light-years, or a hundred times the diameter of the Milky Way.

At the center lies an active galaxy that is thought to contain a black hole with the mass of a billion suns. As the hole sucks in the surrounding gas, the gas is forced to spin into a flat "accretion disk." The disk ejects two jets of relativistic electrons and protons in opposite directions—the long threads we see. The actual structure of the jets is controversial, but they must contain strong ordered magnetic fields that keep the jets aligned. The jets collide with the tenuous intergalactic gas and inflate the huge lobes with energetic charged particles. The whole object glows in radio radiation.

This object, the largest of many like it, contains several of the key magnetic structures we'll be discussing, namely, accretion disks around black holes and protostars, jets, and magnetized bubbles, or "magnetospheres."

Astronomers are currently learning more about cosmic magnetism than ever before. They are using the latest Earth-based optical and radio telescopes and infrared satellites to explore how the fields operate in different settings. Computer simulations of basic magnetic processes have come a long way, too. But this is only the beginning. In the near future we will be able to observe with even more sensitive millimeter-wave interferometers and larger space telescopes. This is an exciting time in the unfolding story of cosmic magnetic fields. I invite you to join me on this voyage of discovery.

We'll begin by recalling how several famous experimenters unveiled the basic properties of magnetism. Then we'll talk about the Earth's field and how studies of the Sun helped to clarify how many astronomical bodies generate magnetic fields. We'll also see how far we've come in simulating these processes with computers.

After that, we'll consider the impressive magnetospheres of Jupiter and Saturn. Star formation is our next topic; we'll see how magnetic fields become critical in the birth of stars and, perhaps, galaxies. We'll take a big jump then to learn how astronomers are able to detect and measure fields in the Milky

Way and in distant galaxies. We'll also examine the enormously powerful fields in some of the densest objects in the universe. Moving out even farther, we'll see how astronomers are mapping the fields between the galaxies and theorizing on how they were formed. Finally, we'll hear the speculations on how the very first magnetic fields in the universe were created.

Hop on board! We have a long trip ahead!

THE MAGNETIC UNIVERSE

GETTING REACQUAINTED
WITH MAGNETISM

Before we plunge into the cosmos, let's review some of the essential properties of magnetism that scientists have discovered in the laboratory over the past two centuries. You're probably familiar with some of them, while other aspects, having to do with lines of force, may be unfamiliar.

Nowadays, a child first encounters magnetism in school while handling a permanent magnet. She sprinkles iron filings on a paper held over the magnet and smiles at the lovely pattern she's created (fig. 1.1). She's taught that these are "lines of force," and she can see that they connect the "poles" of the magnet. With two magnets in hand, she can actually feel the force of repulsion of similar poles and the attraction of opposites. She learns that the Earth has such lines of force and is itself a big magnet.

Until the early nineteenth century, the most advanced scientists knew no more about magnetism than does this modern child. As we'll see in the next chapter, Sir Edmund Halley, of comet fame, had previously mapped the Earth's magnetism as an aid to navigation, but even he had no understanding of the origin of its force.

Hans Christian Oersted, a Danish natural philosopher, made the first critical discovery in 1820. He was born in Denmark in 1777 of poor parents, and as a young man he studied to become a pharmacist. His strong interests in philosophy and science led him to abandon pharmacy, however, and to pursue an academic career. He became intrigued by the possibility that all the forces of

1

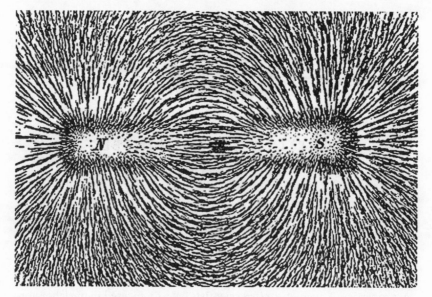

Fig. 1.1. Iron filings around a bar magnet outline lines of force.

nature must originate from a common source. (He was correct! See note 1.1.) In particular, he suspected that the forces of magnetism and electricity were closely related, which was contrary to the current opinion of experts.

Electricity had been fairly well explored by Oersted's time. The ancient Greeks had discovered static electricity by rubbing amber; Charles du Fay, a French scientist, had distinguished between positive and negative electric charges; and the Leyden jar, a device for storing charge, had been invented. Joseph Priestley demonstrated in 1767 that opposite charges attract each other with a force that decreases with the square of their separation. In 1800 Alessandro Volta built the first chemical battery, a reliable source of electric current. The stage was now set for Oersted to make a fundamental discovery.

After quitting his job as a pharmacist, Oersted traveled widely, delivering a series of public lectures on science. In time, he built a reputation that gained him an appointment at the University of Copenhagen, where he gave public demonstrations of chemistry and physics. In 1820 he was preparing to demonstrate how an electrical current heats a wire and produces light. While setting up his equipment, he was struck by the possibility that the current might also affect a nearby compass, but he had no time to test the idea before his lecture.

Oersted was so excited by his idea that he took the bold step of testing it in the midst of his lecture. He placed a compass below the wire. When he switched on the electric current, he noticed, with tremendous satisfaction, that the compass turned from north to align itself perpendicular to the wire. When he reversed the direction of the current, the compass also reversed its direction.

He had discovered two important relations. First, a steady current in a wire creates a magnetic force in its neighborhood. Second, the force points in a definite direction that depends on the direction of the current. As he had long suspected, magnetism and electricity are indeed closely connected. Later, he made a careful study of the effect and wrote up his results for publication.

Other scientists jumped in to extend Oersted's discovery. André-Marie Ampère, a French professor of mathematics two years older than Oersted, was among the first. Ampère had been a child prodigy who mastered much of known mathematics by the age of 18. After he heard of Oersted's discovery, he immediately carried out his own experiments and, within a week, submitted a quantitative theory for publication.

Ampère showed that the magnetic force around a current-carrying wire is always tangent to a circle perpendicular to the wire (fig. 1.2). Moreover, the strength of the force is proportional to the current in the wire and decreases as the distance from the wire increases. (One can use this relation to define the unit of magnetic force, or "flux density," as note 1.2 describes.)

Ampère also wound a current-carrying wire into a helix (a "solenoid") and showed that the resulting magnetic force was constant inside the helix and had the same "iron-filings" pattern outside the helix as a bar magnet. (This result would prove to be important in explaining the field of the Earth.) Finally, he proposed an explanation for the origin of magnetism in a permanent magnet. He suggested that each iron molecule is in effect a tiny solenoid, with a circulating current. When all the little solenoids are forced to point in the same direction, the iron bar becomes a strong permanent magnet.

Ampère's experiments were the first of a new branch of physics, which he named electrodynamics. The common unit of electric current was named in his honor (note 1.2).

These initial discoveries concerned *steady* currents and *steady* magnetic fields. Michael Faraday, a brilliant English experimenter, took the next important step. Faraday is the outstanding example of a self-taught genius who raised himself by his own efforts. Because his parents couldn't afford to send the boy to school, Faraday was apprenticed to a bookbinder. He learned science by reading the books he was given to bind. These included the *Encyclopaedia Britannica*, which had articles on electricity and magnetism by Sir Humphry Davy. He wrote to Davy and obtained a position as his assistant. Gradually he gained the freedom to do his own research, principally in electrodynamics and chemistry.

In 1831, Faraday discovered that when he thrust a bar magnet through a closed loop of wire, he could detect a momentary surge of electric current in the wire. Similarly, a current appeared if he moved the loop past a fixed magnet. It was the *relative* motion that mattered. In fact, any change in the magnetic field near the loop "induced" a current. The size of the current depended on the rate of change; a faster relative motion induced a larger current.

Oersted had shown that a steady current creates a steady magnetic field. Now Faraday had shown that *a changing magnetic field produces a changing current.*

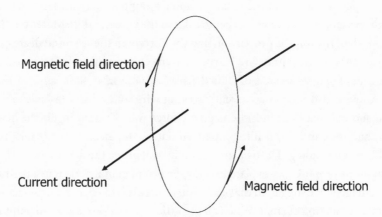

Fig. 1.2. Ampère showed that, at any point near a current-carrying wire, the direction of the magnetic force is always tangent to the imaginary circle whose plane is perpendicular to the wire and that passes through the chosen point. The strength of the magnetic field is constant on the circle and decreases proportionately to the distance from the wire. Note the "right-hand rule": if you grasp the wire with your right hand, your fingers point in the direction of the field vector.

To explain this important result, Faraday invented the concept of lines of force. When one performs the iron-filings experiment, one can imagine lines that connect the filings in curves around the bar magnet (fig. 1.1). The lines crowd together at the poles, where the force is strong, and spread out where the field is weak. So Faraday proposed to represent the strength of the force by the spacing or density of lines of force. He called the number of lines that thread through a unit of area the flux density. His law of induction amounts to saying that the rate of change of the total flux through a loop determines the current in the loop.

Lines of force, I repeat, are *imaginary* aids to visualizing magnetic forces, but since Faraday's time scientists could not do without them. We commonly speak of force *fields* now: the gravitational field, the magnetic field, and other subatomic fields. A field is a region in which a test object, like an apple or a moving electron, experiences a force at every point in space.

An electron or proton that is moving at some angle to a magnetic line of force feels a force (named for Hendrik Lorentz, the famous Dutch physicist) that pulls it into a circle about the line. The particle will end up spiraling around the line indefinitely unless some other forces, like collisions or electric fields, knock it loose.

Faraday went on to invent a machine important to our story, the dynamo, which is the precursor to our modern electric generators. His dynamo consisted of a copper disk that rotates between the poles of a horseshoe magnet (fig. 1.3). As he turned the disk, lines of force sweeping through the disk induced a direct electrical current that passed through sliding brushes to an external circuit. Here, mechanical energy from Faraday's arm converts to electrical energy.

Finally, James Clerk Maxwell appeared on the scene around 1856. Maxwell was a Scottish physicist who combined high mathematical skills with a strong visual imagination. He was able to appreciate Faraday's concept of moving lines of force and to create equations to describe their behavior quantitatively. We won't describe how he arrived at his equations. Suffice to say that over a decade of hard thought, he constructed a set of four compact equations that incorporate the findings of Oersted, Ampère, and Faraday.

His theory led to the astounding prediction that changes in the magnetic field of a source propagate in empty space at the speed of light. That implied

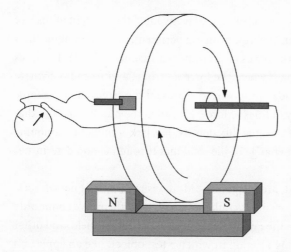

Fig. 1.3. Turning the disk in Faraday's dynamo through the field of the permanent magnet induces a current in the disk that flows between the brushes (the dark patches on the wheel and axle) and through the DC ammeter (on the left). Mechanical energy is converted to electrical energy.

that light itself is an "electromagnetic" wave that has oscillating electric and magnetic fields. When Heinrich Hertz confirmed this prediction in his laboratory (in 1889), and showed that electromagnetic waves have all the properties of light (reflection, refraction, etc.), all of Maxwell's equations were proven valid.

Most astronomers I know don't think in terms of Maxwell's equations when tackling a new problem. Instead they rely on Faraday's lines of force to get a physical picture of what's going on. So lines of force are still very useful. If we assign certain properties to the lines, they can represent all the laws discovered by Oersted, Ampère, and Faraday and incorporated in Maxwell's equations.

Here are the important properties. First, lines of force (LOF) are always closed contours, which may have bends or kinks. That means that isolated north or south poles do not exist. If you saw off the north pole of a magnet, you immediately create a corresponding south pole somewhere on it and the poles are again linked by closed LOF.

Second, LOF act like strong rubber bands that can be stretched indefinitely without breaking. Like rubber they resist being pulled. To stretch LOF you have to pull hard against their tension, which means they store energy. When released they want to snap back and release their stored energy. Like rubber bands they also resist being compressed and can exert a sidewise pressure. In fact, if unconstrained, LOF tend to expand outward indefinitely.

Finally, LOF never cross. But oppositely directed LOF that *touch* can "cancel" and release stored energy. (Solar flares are one striking example of this phenomenon.) On the other hand, LOF with the same direction that touch resist being squeezed together.

The vast spaces between stars are not empty. As we shall see again and again, the spaces contain low-density gases, dust, and magnetic fields. If the gas molecules have lost a sufficient number of electrons by exposure to hot starlight, the gas is called plasma and is a weak conductor of electricity. Therefore, any relative motion between plasma and an existing magnetic field would induce electrical currents, according to Faraday's "law" of induction. This effect has interesting consequences, which Hannes Alfvén, a controversial Swedish astrophysicist, pointed out in 1942.

Alfvén predicted the existence of a novel type of wave in magnetized plasma. When he presented his work at conferences, he was met with much skepticism because Maxwell's equations forbid an electromagnetic wave from penetrating a conductor of electricity. Alfvén later wrote with some bitterness that his prediction was accepted only after Enrico Fermi, the great Italian physicist, heard him speak. Fermi exclaimed, "Of course such waves must exist!" with the implication that Fermi should have discovered them himself. Alfvén waves were eventually produced in the laboratory and are now part of the astrophysical canon. Alfvén shared the Nobel Prize in Physics for his many contributions to the new field of hydromagnetics.

Alfvén waves involve a strange coupling of the magnetic field and the plasma in which it is immersed. The two are "frozen" together, in Alfvén's colorful description. He meant that any relative motion between them would induce electrical currents that would resist their separation. They must move together, even if that requires that the field lines bend or twist.

We can see this effect in the following laboratory experiment, which has actually been done. Imagine a cylinder of liquid sodium metal, a good conductor of electricity. The cylinder is mounted between the poles of a strong magnet, as in figure 1.4. Now the sodium is made to swirl around the cylinder's axis.

What happens? The field lines are anchored at the poles but can bend else-where. The sodium drags them in the direction of rotation and bends them into a hairpin shape. As the rotation continues, the field lines are stretched and forced to wrap around and around the axis. Each rotation adds another line of force next to its neighbor, which increases the effective strength of the field, according to Faraday's rule. At some critical point the tension in the field lines exceeds the ability of the swirling sodium to stretch any further. Then the flow breaks down into turbulence. We'll see something like this happening as new stars form out of inter-stellar gas clouds.

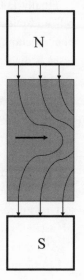

But that's enough of this review. Now it's time to move on.

Fig. 1.4. In this experiment, liquid sodium starts to rotate in a cylinder be-tween the poles of a magnet. The field lines, which were initially straight in the sodium, become bent as the rotation starts. Further rotation will wrap the lines around the axis of the cylinder. This example simulates the "freezing-in" of magnetic field in plasma.

THE EARTH

ON THE NIGHT OF MARCH 6, 1716, Astronomer Royal Sir Edmund Halley was awakened by the noise of a great crowd outside his London home. People in the street were cheering and pointing at the sky. He dressed quickly and stepped into his garden. When he looked up, he was thunderstruck. A luminous cloud, with shimmering rays of yellow, red, and green, stretched from horizon to zenith. He was witnessing the most brilliant aurora of the eighteenth century, the first he had ever seen. He remained watching the spectacle until three in the morning, when the aurora began to fade.

Halley noticed immediately how the rays lined up north to south. They reminded him of the field lines of a magnetized sphere, as revealed by iron filings. He realized he was seeing the magnetic field of the Earth, made visible by some unknown process in the upper air. Figure 2.1 shows a typical example of an aurora.

Auroras are practically the only way we humans can see a magnetic field with the naked eye. We can't sense the field's presence any other way. It is tasteless, odorless, and silent. Birds, dolphins, tuna, and some microbes are better equipped, however. They have special receptors in their bodies that act like little compasses. In effect they possess a sixth sense.

If we were born with such receptors, we might have discovered the Earth's magnetism sooner. As it was, we stumbled across it only two thousand years ago. Since then we have struggled to understand what it is and how it behaves. Only within the last two hundred years have we studied it in the laboratory. Only in the past fifty years have we learned that it shields us from lethal

Fig. 2.1. An example of the northern lights, the aurora, over Oklahoma.

cosmic rays. And only in the past ten years have we tried to simulate the Earth's field in a computer. It's been a long journey.

OF COMPASSES AND MAPS

We really don't know who discovered magnetism, or when. Several conflicting fables do exist, though. As Pliny the Elder, the Roman historian and naturalist, tells the story, an ancient Greek shepherd named Magnes was tending his flock on Mount Ida in Crete. When he walked across a patch of black rock, he felt a strong pull on his iron-shod sandals and on the iron ferrule of his staff. From this episode, if we can believe Pliny, the word "magnetism" was coined.

Thales, a Greek natural philosopher of the sixth century BCE, told a different story. He was quite familiar with such rocks and wrote that they were common in Magnesia, a region in Asia Minor. The problem here is that there were *two* such regions: one in Thessaly, one in Turkey near the river Meander. Nevertheless, the words "magnet" and "magnetism" may have been derived from the name of the region, wherever it was.

As you might guess, the ancient Egyptians also had fables about magnetic rocks, and the Mayans knew about them well before Columbus arrived in the New World. So it's likely that magnetic rocks (later called lodestones) were discovered in many places at different times.

Recent excavations suggest that the ancient Chinese might very well have been first. Archaeologists have found houses built in the Shang Dynasty (1600–1046 BCE) that were aligned with magnetic north. That suggests that the Chinese had made a fundamental discovery at least 2,800 years ago.

A practical people, the Chinese soon applied this discovery to make money. Around 800 BCE, fortunetellers were using the strange force to predict the future. They would spin a spoon-shaped lodestone on a bronze panel that was marked with astrological signs. The fortuneteller knew that the lodestone would tend to point south when it stopped, so he could arrange to have a preselected sign in place. The client was presumably impressed.

Masters of the Taoist art of feng shui also used the lodestone's ability to point south. They set a small lodestone in a piece of wood and floated the wood in a bowl of water. They used these crude compasses to orient buildings and furniture in the most beneficial directions with respect to the invisible currents of energy in the Earth.

Around 200 BCE, someone in China magnetized an iron needle with a lodestone and floated it on a straw. When it settled down, it pointed north and south. The rest is history, as they say.

Chinese merchants were navigating with such compasses as early as the year AD 1000. By 1300 the instrument had spread throughout the Middle East and Europe. Marco Polo and his son may have used one to reach the court of Kublai Khan. Columbus and Vasco da Gama depended on the compass to find their way across the terrifying oceans.

All this voyaging was going on with no understanding of how or why a compass behaves as it does. Not that people didn't speculate. Two popular ideas were that the North Star or possibly a huge mountain of iron attracted the needle of a compass.

Although the compass was an indispensable aid to sailors, it was not especially accurate. Mariners noticed that the compass pointed several degrees off the direction to true north, as determined by the stars, and this difference (named the declination) varied as they sailed around the globe. Also, the compass needle had this puzzling tendency to dip below the horizontal by different amounts as one sailed north or south. This angle is named, logically, the dip.

The behavior of the compass remained a mystery until William Gilbert, physician to Queen Elizabeth I, took up his extensive studies of electricity and magnetism, around 1600. Every schoolchild learns about Galileo, but few adults have even heard of Gilbert.

He was one of the first experimental physicists, who performed careful tests of clear hypotheses. He described how to create an artificial magnet by stroking a bar of iron with a lodestone and how heating a compass would destroy its properties. He disproved the popular superstition that garlic destroys a compass's power. He also investigated the force of attraction between electrically charged objects and gave the name "electricity" to the invisible substance removed by rubbing amber, which is called *elektron* in Greek.

Gilbert suspected that compasses behave the way they do because the Earth itself is a giant lodestone. To test his idea, he shaped a lodestone into a sphere (a terrella) and explored its magnetic properties with a small compass. He demonstrated how, at any longitude, the compass points to a pole, and that the dip of the needle below the horizontal varies with latitude, all in accord with the experience of the navigators. In his monumental treatise *De Magnete*, Gilbert concluded that the Earth is indeed a giant magnet.

He was well ahead of his time, but he was not infallible. For example, he thought the magnetic poles coincide with the geographic poles, and that the compass's deviation from geographic north (the declination) is caused by the attraction of the nearest continent. On the other hand, he proposed to correct maps for the declination in order to improve navigation, and he also proposed to determine one's latitude by measuring the dip of the compass needle. Exactly at the north magnetic pole, for instance, the needle would point straight down.

In proposing these schemes, Gilbert assumed that the Earth's magnetism was fixed for all time. Unfortunately, that is not so. In 1635, the English astronomer Henry Gellibrand discovered that the declination at London had changed by a significant amount in only twelve years. What was worse, the changes were later found to differ at various locations around the world. (According to Joseph Needham, eminent scholar of Chinese science, the Chinese had discovered such changes around 720.)

Such a threat to precise navigation was intolerable. Therefore, the British Admiralty turned to Edmund Halley, the Astronomer Royal, and asked what could be done. Halley is best known for his recurring comet, but he also discovered that stars seem to move relative to their neighbors, and he suggested that transits of Venus across the disk of the Sun could be used to determine the distance to the Sun.

He was a man of many parts, an archaeologist as well as a controller of the Mint. Perhaps his greatest contribution to science, however, was his role in helping Isaac Newton. When the Royal Society balked at the cost of publishing the *Principia,* in which Newton derived for the first time the elliptical orbits of planets from the law of gravitation, Halley stepped in and paid for it out of his own pocket.

As Halley studied the Admiralty's problem, he saw no way to predict the pesky variations in declination. He proposed instead to measure them all around the sea-lanes. They would have to be remeasured every decade or so.

With this objective in mind, Halley was commissioned in 1698 by the Admiralty to conduct a magnetic survey of the Atlantic Ocean and its borders. His voyage took him to the coast of Brazil and far south toward Antarctica. He came home, plotted his data, and connected the points of equal declination with curved lines. His map of 1702 was the first of its kind, and its revisions guided the British fleet for a hundred years.

As data accumulated, Halley noticed how the line of zero declination drifted westward. Apparently the whole pattern of Earth's magnetism was creeping toward the west at a pace that would complete a circuit in a couple thousand years. To explain this curiosity, Halley proposed that the Earth has two north and two south magnetic poles. Two of these poles lay on the surface, the other two lay on an inner sphere about 800 km deep. By cleverly adjusting the slow westerly rotation of this inner sphere, he could reproduce the observed variations to some extent.

Halley's concept of four poles was attacked, but it persisted into the nineteenth century, as scientists attempted to account for the small deviations of the field from a perfect dipole. It was discarded after Carl Friedrich Gauss introduced the modern method of describing the Earth's field.

Gauss was a child prodigy. In a famous story, he was asked by his teacher to find the sum of all the numbers from one to a hundred. Instead of tediously adding them all, he instantly discovered that they could be paired: 1 plus 100 equals 101, 2 plus 99 equals 101, and so forth. He easily obtained the sum as 101 times the number of pairs, 50. He rapidly developed into one of the foremost mathematicians of the nineteenth century, who made important discoveries in number theory, differential geometry, topology, and the error analysis of observations. He also contributed to such diverse fields as astronomy, geodesy, electricity, and optics.

Gauss spent most of his career as a professor of mathematics at the University of Göttingen. He became interested in the mathematical problem of describing the Earth's magnetic field quite early in his career but put the work aside for lack of sufficient observations. In 1828, however, he spent three weeks at the Berlin home of Alexander von Humboldt, one of the greatest scientists of the age. Humboldt had been interested in geomagnetism for decades and had organized a network of magnetic stations around the world to collect observations. He tried to interest Gauss in moving to Berlin and joining him in the effort to analyze the data, but Gauss could not be tempted and returned instead to his comfortable berth at Göttingen.

Nevertheless, his interest in geomagnetism had been reawakened. In Berlin, he had met Wilhelm Weber, a brilliant young physicist, whom he recommended for a professorship at Göttingen. When Weber arrived at the university, the two men became close friends and co-workers.

They embarked on a series of remarkable experiments on the connection between electrical currents and magnetic forces. They also invented a method to measure the "absolute" strength of a field and to express the result in units of length, mass, and time. In the course of their work, Weber demonstrated a crude telegraph system several years before Samuel Morse. Gauss tried to interest the military in the invention but was unsuccessful. Gauss and Weber did succeed, though, in organizing an informal network of magnetic stations in Europe to acquire more observations of the terrestrial field. This Magnetic Union produced a preliminary map within six years.

In 1833 Gauss published his solution to the problem of describing the Earth's magnetic field. He showed how to combine observations of the field from many stations. The global field could be represented as a weighted sum of elementary shapes. The basic shape is the "dipole," the familiar pattern around a bar magnet, with two magnetic poles. The next simplest shape, the "quadrupole," has four poles, two north and two south. Add the two shapes in the optimum proportion and orientation, and you can fit the data better than you could with the dipole alone. Add more shapes, the "octupole" and so on, until you have fitted the data as well as possible. Current mathematical models of the Earth's field employ as many as 168 "multipoles" to achieve the required accuracy. Gauss's method has an additional benefit; it allows one to reconstruct the shape of the field *inside* the Earth from observations at the surface.

Although Gauss was able to estimate the location of the South Magnetic Pole, the precise location of the North Magnetic Pole was uncertain. So a milestone was reached in 1831, when James Clark Ross discovered the position of this pole. Ross was first mate on an expedition to find the Northwest Passage of mythical fame, a shorter route to the Far East. When his ship froze into the ice pack off northern Canada, he took the opportunity to march over the ice to find a place where the compass needle pointed straight down. The pole lay at a latitude of 70 degrees, 5 minutes, nearly 2,000 km from the geographic pole (note 2.1).

By the early nineteenth century, the errors in existing declination maps were no longer tolerable. The only solution was to remeasure the field at many locations around the globe. So around 1830 an international group of scientists, led by Humboldt, Gauss, and Weber, petitioned the British Admiralty to establish a worldwide network of geomagnetic stations.

This Magnetic Crusade was approved in 1839. Observatories were set up around the British Empire, at Greenwich, St. Helena, Cape Town, Tasmania, and Toronto. Later, Melbourne and Bombay were added. Oslo was the first of many foreign cities to join the campaign.

Sir Edward Sabine, an Irish astronomer, was put in charge of this immense cooperative effort. He was known primarily for his precise measurements of the Earth's shape, using pendulums. He had also made magnetic measurements during two Arctic expeditions in search of the Northwest Passage. He noticed that the magnetic field strength had changed in the interval between voyages and guessed that the poles might have shifted.

Sabine proved to be an excellent scientific director. He continued in this role for the next thirty years, collecting data from the many stations and updating maps. While engaged in this work, he learned that the network of stations occasionally recorded violent fluctuations of the field strength that lasted for several days. Sabine realized that the number of magnetic disturbances rose and fell with a period of about eleven years. He knew from the researches of Heinrich Schwabe, a tenacious German astronomer, that the number of sunspots varies in an eleven-year cycle. Putting these two facts together, he proposed that the Sun has an important effect on the Earth's field. The causes of its effect remained unknown, however.

In 1904, Roald Amundsen (who later would be the first to reach the South Geographic Pole) was fighting his way through Canada's Northwest Passage. On April 26 he located the North Magnetic Pole and determined that it had moved about 30 km since Ross had found it in 1831. Sabine's suspicions were proven correct. The pole has accelerated since then, moving northward at an average rate of 10 km/yr. At its present rate of 40 km/yr it could leave Canada soon and head for Siberia. The South Magnetic Pole is also wandering, independently of its northern mate.

In the early years of the twentieth century, two French scientists discovered perhaps the most astounding feature of the Earth's magnetic field. These "paleomagnetists" were studying the fields embedded in old lava fields. Lava contains magnetite, and as the lava cools the mineral retains the direction and strength of the Earth's magnetic field. Bernard Brunhes and P. David found in 1906 that the field direction in some lava rocks was opposite to the present field. Instead of pointing toward the North Pole, their fields pointed south. They speculated that the Earth's global dipole field might have *reversed* in the past.

The plot thickened with a report in 1926 that rocks with reversed polarity were found all over the world. There were clues that some reversed rocks were older than others, but dating methods were primitive. Only after the potassium-argon method of dating rocks was invented was the final picture revealed: the Earth's poles have reversed polarity at random times in the past. During the last 10 million years, the average interval between reversals is 250,000 years, but the last reversal occurred 780,000 years ago. In one well-documented reversal, the intensity of the field decreased by 10 percent, while its direction swung from north to south in only a few thousand years.

These reversals of the global field were crucial to a revolution in the science of geology. In 1963, Frederick Vine and Drummond Matthews were measuring the magnetism of rocks on the floor of the Atlantic Ocean, in the vicinity of the mid-Atlantic submarine ridge. This ridge is part of a chain of submarine mountains 50,000 km long that winds around the globe like the stitching on a baseball. They discovered a pattern of zebra-like stripes, parallel to the ridge, in which the magnetic polarities of the stripes alternated from normal to reverse.

They proposed that molten lava spews from the ridge and flows down its sides. As the lava cools, the magnetite in the basalt "freezes" the existing magnetic field. Dating the reversals of polarity demonstrated that the stripes are arranged in order of age; the farthest from the ridge were oldest. And that implied that the seafloor was continually spreading out in both directions from the ridge at a rate of a few centimeters per year.

The idea of seafloor spreading was largely the brainchild of Harry Hess, professor of geology at Princeton University. As early as 1959 he had proposed that the seafloor was part of a system of gigantic tectonic plates that were constantly growing at their edges and submerging in deep trenches at the coasts of continents. His ideas were the basis of the grand revision of the history of the Earth's crust, the theory of plate tectonics.

Geomagnetic stations around the world have continued to pump out daily measurements of the field strength and direction and its variations up to the present time. Most of the ground-based stations are located in Europe, and coverage in Africa, Siberia, and South America is particularly sparse. Observations from ships and aircraft are useful but intermittent. Satellites have therefore become essential tools. Since the 1960s a series of them have been launched to measure the vector field. Most recently, the Danish Oersted satellite was launched in 1999 and orbits at a height of around 700 km. Germany's CHAMP has been returning data since 2000 from an orbit at about 400 km.

From these streams of data, mathematical models of the global field are continually updated at several centers around the world. Two examples are the World Magnetic Model and the International Geomagnetic Reference Field. The WMM is the standard for navigation in the United States and the United Kingdom; the IGRF is used throughout the world for research.

We can get a broad overview of the Earth's present field from the WMM. Figure 2.2A shows the horizontal component of the field, measured in nanoteslas (billionths of a tesla or hundred-thousandths of a gauss). Please notice that the lines in the figure are not lines of force but contours of constant horizontal field. A perfect dipole would have contours running parallel to the equator, with the strongest field at the equator.

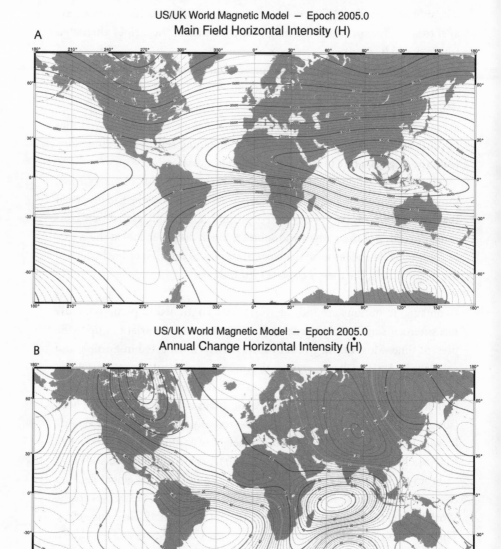

US/UK World Magnetic Model – Epoch 2005.0
Main Field Horizontal Intensity (H)

A

US/UK World Magnetic Model – Epoch 2005.0
Annual Change Horizontal Intensity (Ḣ)

B

Fig. 2.2. Shown here are the contours of the horizontal component of the measured field of the Earth, in nanoteslas (tens of microgauss) from the World Magnetic Map of 2005.

The measured field is predominantly dipolar, as you might expect, but its axis is tilted by about 10 degrees from the Earth's rotational axis. Notice the South Magnetic Pole in Antarctica, indicated by closed contours. A secondary south pole appears in the South Atlantic and a secondary north pole over Malaysia. So the field has an obvious quadrupolar component.

Figure 2.2B displays projected rates of change of the horizontal field for the coming decade. The fastest decrease is occurring over the South Atlantic; the fastest increases are over the Indian Ocean and northern Canada. Nobody knows how to predict such changes from first principles.

From 1800 to the present, the dipolar component of the field has weakened at 6.3 percent per century and the quadrupolar component has strengthened. Measurements of the magnetism of rocks indicate that the dipole's decrease has persisted for at least 2,000 years. If it continued to decrease at the present rate, it would vanish in 1,600 years.

Let's summarize and ask what a satisfactory theory must explain. First and foremost is the very existence of the field. How does the Earth generate its field? Why is it multipolar? And how can we account for its variations, both long- (secular) and short-term? What causes the poles to be displaced from the rotation poles by as much as 2,000 km and wander independently by as much as 40 km/yr? Why does the whole field drift slowly to the west? Most puzzling of all is the reversal of polarity every few hundred thousand years. Here is a challenging set of problems for geophysicists to solve.

THE ROAD TO THE DYNAMO

Before scientists could begin to explain the origin of the Earth's field, they needed some crucial laboratory results to establish a connection between electrical currents and magnetic fields. In the previous chapter we recalled some of the relevant experiments. In 1820, the Danish scientist Hans Oersted discovered that a steady electrical current flowing in a wire loop generates a steady magnetic field around the loop. Then André-Marie Ampère showed that if several loops are wound in a helix, the field lines are straight and parallel inside the helix and have a dipole shape outside the helix. That result suggested to physicists that internal electrical currents might generate the Earth's dipole field. But how are the currents created?

Michael Faraday, a talented British experimenter, had found a possible answer. He had shown in the 1830s that moving a bar magnet through a loop produced a momentary current in the loop. Moving the loop past the magnet worked equally well; what mattered was the *relative* motion of a field and a conductor. Faraday went on to demonstrate how to create a steady current with a device he called a dynamo.

In Faraday's dynamo (see fig. 1.3), a disk of copper is free to rotate in the field of a permanent magnet. When Faraday cranked the disk, a steady current flowed in the circuit. The faster the speed of rotation, the stronger the current. The dynamo was converting mechanical energy (from the experimenter's hand) to electrical and magnetic energy. In a more elaborate version of a dynamo, the current that it generates is used in part to create the magnetic field that it requires. In other words one doesn't need a permanent magnet to supply an external field. (Today we rely on this form of Faraday's dynamo, an "alternator," to generate electricity for our homes and factories.)

Sir Joseph Larmor, an eminent Irish scientist, was the first to apply Faraday's dynamo toward explaining the origin of cosmic magnetic fields. Larmor was famous for his research on the alignment of atoms in a magnetic field and for an electron theory of the atom. Like Hendrik Lorentz and George FitzGerald, he independently discovered the formulas of special relativity before Einstein explained their physical significance.

In 1919, Larmor was attempting to explain the Sun's magnetic fields. George Ellery Hale, director of the Mount Wilson Observatory, had recently discovered magnetic fields in sunspots as strong as 3,000 gauss. He had determined that a sunspot's field lines emerge as a vertical column at the Sun's surface and spread out at higher altitudes, like a tree trunk and its branches.

Larmor proposed that, at the Sun's surface, differences of temperature drive electrically conducting gas to flow horizontally into the vertical sunspot field and out at higher altitude. As the gas crosses the field lines, it induces a current that flows around the vertical axis of the sunspot. This current, Larmor suggested, might be the actual source of the magnetic field of the sunspot. As long as a source of heat was available to drive the horizontal flows, the system would act as a natural dynamo.

Notice that Larmor's dynamo is "self-exciting"; it generates the same magnetic field that it needs to function.

In the 1930s, geophysicists hoped to apply Larmor's ideas to explain the Earth's field. One could imagine a kind of self-excited dynamo in the Earth's interior. Perhaps hot magma conducts electricity well enough to act like the copper disk in Faraday's dynamo. As the magma rotates through the dipolar magnetic field, it would generate electric currents. And these currents would in turn generate additional magnetic fields that would maintain the existing field. This may sound like some sort of perpetual motion machine, but remember that the energy to maintain the currents and the field would ultimately derive from the energy of the flow of lava.

In 1933, however, British astrophysicist Thomas Cowling proved a little theorem that torpedoed Larmor's neat scheme. He showed that *no type of motion* that is symmetric around an axis could maintain a dynamo. In particular, no variations of rotation speed either in depth or along the Earth's axis of rotation could be effective. His proof was aimed at Larmor's model of sunspots, but it applies universally. It was devastating.

Progress in explaining the Earth's field stalled until geophysicists learned much more about the interior of the Earth.

A SMATTERING OF SEISMOLOGY

Sir Isaac Newton made the first important discovery about the Earth's interior. Using his studies of the force of gravity, he was able to calculate that the average density of the Earth is more than twice that of the rocks at the surface. That meant that the Earth has a dense core.

Seismologists, who study the waves that earthquakes produce, eventually confirmed the existence of a core and gave us good estimates of its size and composition. To follow their reasoning, we must digress slightly and talk about the waves that earthquakes produce.

Rocks transmit two kinds of waves. In pressure (P) waves, the particles of rock oscillate *along* the direction in which the wave is traveling. (Sound waves in air are P waves.) In shear (S) waves, the particles oscillate *perpendicular* to the wave's direction. Solid rock can transmit both types, but molten rock can transmit only the P wave, because two sliding layers can couple only weakly. Both types of waves are reflected and refracted as they pass between layers of different temperature and composition.

Richard D. Oldham, an Irish geologist, was the first to identify such waves. In 1906 he noticed a "shadow zone" on the hemisphere opposite an earthquake, in which P waves were excluded. The zone had the shape of a wide ring that extends from 105 to 142 degrees from the location of the earthquake.

Oldham attributed this shadow to the refraction (or bending) of P waves through a dense core. He pictured the core as acting as a spherical lens for P waves, in just the same way that a glass ball would refract light. From the angular limits of the shadow zone, he was able to estimate the diameter of the core as about 7,000 km. Because P waves traveled more slowly through the core, he concluded that it must be denser than its surroundings, but he couldn't say more about its composition.

The next important step was made in 1926. The German geophysicist Beno Gutenberg discovered a shadow zone for S waves. It covers almost the entire hemisphere opposite an earthquake, starting at an angular distance of 105 degrees. Gutenberg concluded that the core must be *liquid*, because S waves cannot travel through a liquid. He calculated the diameter of the core as 7,700 km.

The picture became still more complicated in 1936. Inge Lehmann, a Danish seismologist, used much-improved seismographs to discover very weak P waves within the P wave shadow. This could only happen, she argued, if the core consisted of a *solid* inner sphere surrounded by a *liquid* outer shell, each with different compositions. From more seismic data, we now know the diameters of the inner and outer cores are approximately 2,500 km and 7,500 km, respectively. So the core is about the size of the Moon and has the temperature of the Sun's surface.

The final pieces of the puzzle came together in the 1940s, as Francis Birch, a physicist working at Harvard University, measured the elastic and thermal properties of many kinds of rocks under high temperature and pressure. He concluded that the solid inner core is probably composed of nearly pure iron, while the liquid outer core is an alloy of iron and nickel, with traces of silicon and sulphur. The iron presumably sank to the core during an early stage in the formation of the Earth.

To complete our picture of the Earth's interior, we note that the region between the outer core and the thin crust is the mantle, which consists of rock in a plastic or semimolten state.

ON THE ORIGIN OF THE EARTH'S FIELD

By the 1940s, geophysicists had sufficient information about the interior of the Earth to begin to make real progress on the origin of its magnetic field. Some form of dynamo seemed the most attractive possibility, but Cowling's theorem of 1933 had blocked the simplest proposals. New physics was needed to attack the problem.

Hannes Alfvén supplied part of the solution in 1942 with his concept of "frozen-in" field lines. When a magnetic field is immersed in a very good conductor of electricity, which may be either a gas or a fluid, the field and fluid must move together. Any relative motion induces strong electrical currents that resist the motion.

Frozen-in fields changed the problem of the origin of the Earth's field. Scientists working on the problem recognized that a highly conducting core could not simply rotate through a stationary field, as Larmor had proposed in his theory of sunspots. Instead the core's rotation would wrap an initial north-south field into a belt of lines parallel to the equator, a "toroidal" field. (Recall that the lines behave like rubber bands that can stretch indefinitely without breaking.) When this toroidal field reached a critical strength, it would begin to slip through the liquid core. That is, the frozen-in condition would eventually break down if the core had a finite conductivity. An equilibrium toroidal field would be the final result.

But the observed field is primarily a "dipole," with field lines stretching between the north and south poles, not a toroidal field. How then is the dipole field created?

Walter Elsasser, a German-born geophysicist, attacked this problem in a series of elegant papers beginning around 1946. He set up equations to describe how the rotation of a liquid metallic core would couple with an initial dipole field. First he showed that the electrical resistance of the core would cause the Earth's dipole to decay in only a few tens of thousands of years unless some motions of the core maintained it. When he turned on the rotation of the core, he confirmed that a toroidal field would develop in it, parallel to the equator. The field would have opposite directions above and below the equator because of the way the field wraps around the globe (fig. 2.3A). Despite all his efforts, however, he could not find additional flows in the core that would generate a dipolar field. Cowling had won out again.

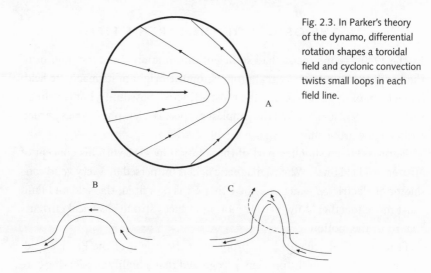

Fig. 2.3. In Parker's theory of the dynamo, differential rotation shapes a toroidal field and cyclonic convection twists small loops in each field line.

Sir Edward Bullard, a senior British geophysicist, added the essential ingredient to the solution of the problem in 1950. Bullard suggested that heat from the Earth's inner core is carried toward the surface of the core by the convection of cells of fluid metal (note 2.2).

Bullard envisioned that such electrically conducting cells could interact with a toroidal field within the core to produce the dipole we observe at the surface. His essential point was that such motions were not necessarily symmetric about the rotation axis and therefore were not subject to Cowling's objections. But he had no proof of his assertions.

In the 1950s, Walter Elsasser held a faculty position at the University of Utah. There he met a young colleague, Eugene Parker, who was intensely interested in the problem of terrestrial magnetism. The abilities of the two men were complementary. While Elsasser attacked a problem with formal mathematics, Parker relied more on his powerful physical intuition. They spent many hours discussing Bullard's ideas.

Today Parker is recognized as one of the two or three most productive and creative theoretical astrophysicists of his generation. He can pose the single critical question that probes the heart of a problem. As we shall see he has contributed many important ideas and insights to the subject of cosmical magnetism. In fact he wrote a book on the subject that remains a classic, *Cosmical Magnetic Fields: Their Origin and Their Activity* (1979).

He's always been a fierce advocate for solar physics, recognizing the Sun as a unique astrophysical laboratory, and has helped to advance many important solar projects. A modest man in all but research, he's one of the few luminaries I know who answers his own mail in longhand. Now over 80, he still maintains his wiry figure by daily walks at 4 miles an hour.

Parker's career took off in 1955. In that year he published a mathematical demonstration of an idea that cracked the problem of the geomagnetic field. In his theory he postulated two belts of toroidal field in the Earth's core, with opposite polarities north and south of the equator, such as those Elsasser had demonstrated. In his example the field points to the east in the northern hemisphere.

He then proposed that convection cells in the core move the same way as "cyclones" in the Earth's atmosphere. In a cyclone, a column of hot air expands as it rises. Because the Earth rotates, Coriolis forces act on the expanding air, causing it to rotate clockwise in the northern hemisphere and anticlockwise in the southern hemisphere (note 2.3).

Parker suggested that a rising column of fluid metal in the core would push up a portion of its toroidal field, forming an omega-shaped loop of magnetic field (fig. 2.3B). In the northern hemisphere, the Coriolis force would turn an expanding loop *clockwise* so that it would lie nearly parallel to a meridian (fig. 2.3C). The field at the top of the loop would then point toward the north. In the southern hemisphere, the toroidal field points west, but the Coriolis force would turn a loop anticlockwise. So the top of that loop would also point to the north.

Then Parker showed how many similar loops in both hemispheres could merge. The fields at the nearly vertical sides of the loops would cancel, leaving only the horizontal component of the field that points to the north—in other words a dipole field.

Parker's concept of cyclonic convection was the key to a successful model of a self-sustaining dynamo. But it was only a sketch of a theory. For one thing, he drastically simplified the formation of loops by ignoring turbulence. One might expect turbulent motions in the convection zone to shred loops before they could join into a coherent poloidal field. Parker would return to this problem much later.

Turbulence is one of those subjects in classical physics that has never been thoroughly understood. In a viscous fluid, small-scale turbulent flow changes so rapidly in time and in space that physicists are totally frustrated in trying to describe it (note 2.4). As a result, physicists have been forced to make strong assumptions in order to make any progress in developing a theory. The problem grows worse when one adds magnetic fields.

Max Steenbeck, Fritz Krause, and Karl-Heinz Rädler, a trio of German physicists, accepted the challenge of constructing a mathematical model of a turbulent dynamo. Around 1964, they invented a method (mean-field magnetohydrodynamics) in which the large-scale motions of rotation and cyclonic convection were specified in advance and small-scale turbulent motions were allowed to interact with the larger flows. By suitably averaging the results, they could investigate the conditions required for a dynamo to be sustained. Later they applied their methods to the Sun and stars, as we shall see.

In the 1970s and 1980s, a deluge of complicated mathematical models of dynamos were proposed for the Earth and Sun. But all of these were "kinematic" models, in which the large-scale ordered motions were specified arbitrarily. Some advanced models were nonlinear, in that they included the reaction of strong magnetic fields on the basic flows. Such models were inherently incapable of predicting the strength of the dynamo fields, but they were able to reproduce some of the long-term secular variations of the Earth's field.

A truly satisfying theory of the Earth's field, however, would include a mathematical description of how convection cells arise by the flow of heat from the interior, how they evolve in time and space, and how they interact with existing magnetic fields to create and maintain a dynamo. If one were to allow turbulence to enter this problem, it could become nearly insoluble. But with the development of powerful supercomputers, starting in the 1980s, physicists began to calculate the behavior of fairly realistic dynamos, in three dimensions and time. Gary Glatzmaier was a pioneer in this challenging research and still is near the head of the pack.

THE LATEST IN COMPUTER MODELS

Glatzmaier was a fluid dynamicist at the Los Alamos National Laboratory for many years. There he was able to use some of the fastest computers in the world. He was interested in such problems as convection in the Earth, the Sun, and some stars; the internal rotation of the Sun; the circulation of the Earth's atmosphere; and the evolution of planets. Gradually he developed computer codes that could follow the evolution of a realistic three-dimensional system, like the Earth. In 1998 he moved to the University of California at Santa Cruz, where he leads a small research group.

His basic method is to set up the mathematical equations that govern the flows of heat, fluids, and magnetic fields in a spherical body. Then, for each step in time, the computer calculates the change of the existing status of the system and updates all the physical variables.

Glatzmaier developed a huge computer code that incorporates the most detailed information available on the rotation, composition, temperature, and density of the core. The code is so big and so complex that it can only run on 256 processors working in parallel. Even so, a typical calculation can take six months to run.

In the 1990s Glatzmaier was collaborating with Paul Roberts, originally from Newcastle University at Newcastle upon Tyne and lately at UCLA. As they gained experience with the code, they were able to improve its performance, so that it could follow the behavior of a dynamo for simulated times of hundreds of thousands of years. Because their model includes much of the essential physics, they were able to calculate both the shape and strength of the field.

In 1995, they announced that their program had run for a simulated time of 40,000 years. They observed that the dipole field shrank in strength and recovered several times. Then, 36,000 years into the simulation, the dipole actually *reversed polarity* within a mere 1,200 years. They had, for the first time, reproduced an event similar to those recorded in magnetic rocks.

As they examined their results, they learned that convection in the outer liquid core continually tries to reverse the polarity of the dipole but the solid inner core prevents a reversal most of the time. Only rarely does a full reversal occur, and even then the field of the Earth never entirely vanishes. For a relatively short time, the polarity of the inner core is opposite to that of the outer core. It takes a couple of thousand years for the fluid flows to reverse the polarity of the inner core.

More recently Glatzmaier and Roberts ran the code for a simulated time of 500,000 years. In figure 2.4 we see snapshots of the simulated magnetic field before and in the middle of a reversal. With such distortions, it's no wonder that we observe the poles to wander! In this calculation they were also able to reproduce the westward drift of the nondipole features of the field, in good agreement with the observed rate of 0.2 degrees per year.

Such successful matches with observations are encouraging, but that is not the same as understanding how the different parts of this complicated machine work. Glatzmaier and his collaborators will have a delightful time over the next few years digging into the details of their calculations to gain the understanding they desire. As we shall see in the next chapter, Glatzmaier and friends are also working hard to understand the sunspot cycle. They have plans to model the planets and stars, as well as volcanic eruptions.

Glatzmaier's results are by no means the last word on the subject. Like everyone else who does these simulations, he had to compromise in representing

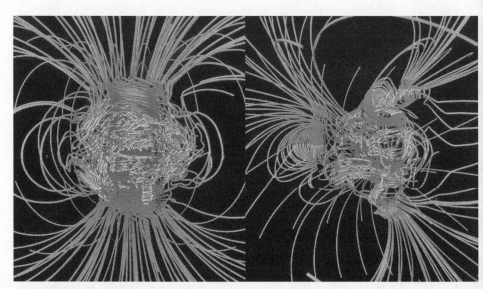

Fig. 2.4. A simulation by Gary A. Glatzmaier and Paul H. Roberts of a reversal of the Earth's magnetic field. The left panel shows the field lines before, and the right panel shows them at the middle of the reversal.

turbulence in the computer. Turbulence is an essential factor in transferring heat and in tangling magnetic field lines. The problem is that turbulent eddies in the liquid core are predicted to come in all sizes, ranging from the thickness of the liquid shell (2,200 km) all the way down to a few meters. No computer has the capacity to follow the full range of sizes in a simulation. Glatzmaier avoided this problem by adopting a rather large viscosity for the liquid iron, which suppresses the smallest turbulent eddies. In effect he assumed the fluid iron to be sticky, like heavy oil.

A Japanese group was able to use a more realistic viscosity in their calculations of the Earth's field. Their powerful supercomputer, the Earth Simulator, ran for thousands of hours to follow the field's evolution over 200,000 years. In July 2005, they reported that they too had witnessed reversals of the field's polarity, but at an average interval of only 5,000 years, not the hundreds of thousands of years determined by dating the seafloor. So it appears that other factors besides turbulence must be tuned to arrive at the observed result. It will take some time for the different groups to sort out these conflicting results and decide how to proceed, but the future of such simulations looks very promising.

At the same time, a dozen different groups around the world are taking an experimental approach toward understanding how the Earth's dynamo works. They are building laboratory models of the liquid core, with internal sensors to measure all the relevant variables.

At the University of Wisconsin, Cary Forest and his group have built a stainless steel sphere 1 meter in diameter, filled with liquid sodium. Sodium conducts electricity easily and has about the same viscosity at 100 degrees Celsius as liquid iron, about that of water. The sphere is immersed in an external magnetic field, which represents the toroidal field of the Earth's inner core. Instead of spinning the sphere, Forest stirs the sodium with two counterrotating propellers. This mechanical arrangement may seem somewhat artificial, but the physicists are trying to simulate turbulent convection in the core without having to boil the sodium.

They carefully designed the machinery to produce only axi-symmetric, large-scale flows in the sodium. According to Cowling's theorem, these flows alone cannot generate a dipole field. So, if they do detect a dipole field, it has to be generated by the turbulent flows interacting with the steady external field.

In February 2006, Forest and friends reported an initial success: they were able to produce a dipole field in the sphere. Now they plan to map the flows in fine detail in order to understand just how the turbulence behaves. Stay tuned!

SUNSPOTS AND THE SOLAR CYCLE

Anyone who has seen recent images of the Sun (fig. 3.1) obtained by such satellites as the Solar and Heliospheric Observatory (SOHO) and the Transition Region and Coronal Explorer (TRACE) can't help being impressed. Far from being the boring round ball one sees at sunset, the Sun is festooned with intricate decorations. Its surface is covered with a forest of x-ray loops and coils. Its outer atmosphere (the corona) is molded into enormous tapered petals and rays. And occasionally, a huge coil will erupt from the corona and float out into space. All these features are signatures of magnetic fields.

Less obvious signs of fields are also present, if you know what to look for. Sunspots are dark because of the powerful magnetic fields they contain. Also, the dramatic flares that explode several times a day derive their brilliant light from magnetic fields. The Sun is truly a magnetic star.

And yet a century ago none of this was known, even to the most erudite scientist. We had to wait for the appropriate technology to confirm what some suspected and to be able to measure what we saw. In this chapter we'll follow the story of magnetic fields on the Sun. It began one summer, long ago, in California.

On June 26, 1908, George Ellery Hale, director of the Mount Wilson Observatory, saw his colleague Charles Abbot walking toward the observatory's dining hall. He hurried to catch up. "With mischief in his eye and almost in a stage whisper he exclaimed breathlessly, 'I think I've got it!!'" Indeed he had (note 3.1).

31

Fig. 3.1. In this TRACE image of the solar corona, dated June 11, 2006, magnetic loops filled with plasma are shining in a spectral line of iron, at a temperature of about 1 million degrees. One edge of the Sun is visible in the upper right. (TRACE is a mission of the Stanford-Lockheed Institute for Space Research, a joint program of the Lockheed Martin Advanced Technology Center's Solar and Astrophysics Laboratory and Stanford's Solar Observatories Group, and part of the NASA Small Explorer program.)

Hale had been working feverishly during the past weeks to test his idea that sunspots contain magnetic fields. He had gotten the idea years earlier, when he was able to photograph the swirling "vortex" of hydrogen gas that lies above a visible sunspot. The vortices seemed to resemble tornados, with a narrow funnel that reached down into the dark sunspot. If these solar tornados were composed of electrically charged particles, he thought, they might generate a magnetic field. How could he look for such a field? In thinking about the problem, he recalled the experiments of the Dutch physicist Pieter Zeeman.

Back in 1896, Zeeman had discovered that a strong magnetic field modifies the wavelengths at which atoms radiate. A hot vapor of iron, for example, normally radiates a unique pattern of sharp bright "lines" in its spectrum. Each line is a narrow range of wavelengths. When the vapor was immersed in a strong magnetic field, Zeeman found that each line splits into three or more lines. The wavelength separation of these components was a direct measure of the strength of the field. Moreover, each component was polarized in a specific way (note 3.2).

Hale realized that Zeeman's discovery was the ideal tool to test his hypothesis about magnetism in sunspots. He began an intensive campaign to photograph the spectra of sunspots, using the newly installed spectrograph at the base of his 60-foot solar tower telescope. On the superb negatives he obtained, he could easily see the splitting of spectral lines that Zeeman had described and that Hale's colleague Arthur S. King had measured in the Pasadena laboratory of the observatory. Later, he was able to verify the polarization patterns that Zeeman reported. There was no doubt about it: the dark centers of sunspots contain magnetic fields of a few thousand gauss, or thousands of times stronger than the Earth's.

Hale's discovery was one of the most important in his illustrious career. And yet there is an ironic footnote to the story. The strong fields of sunspots have nothing to do with the vortices that started Hale on his quest. In fact few sunspots exhibit well-defined whorls, and those that do have an entirely different origin.

Nevertheless, Hale's discovery opened a vast new topic in astrophysics. Dozens of questions were raised. For instance, how could a prosaic ball of hot gas like the Sun generate such strong magnetic fields? Does the Sun have a global dipole field like the Earth's? Could the Sun's magnetic field affect the Earth somehow?

Hale, a great reader, was aware of two remarkable regularities that govern the behavior of sunspots. Heinrich Schwabe, a German pharmacist and amateur astronomer, had counted the daily number of sunspots for seventeen years and reported in 1843 that they wax and wane in an eleven-year cycle. Twenty years later, Richard Carrington, a British amateur astronomer, announced that sunspots emerge in a definite pattern. Instead of popping up at random places at random times, spots are born closer and closer to the Sun's equator as the cycle progresses (fig. 3.2). With the discovery of sunspot magnetic fields, these regularities took on new meaning.

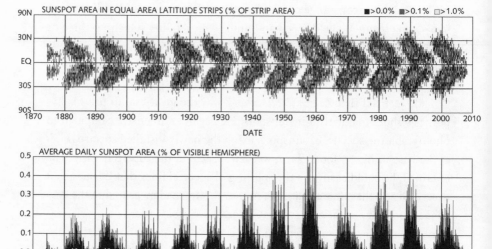

DAILY SUNSPOT AREA AVERAGED OVER INDIVIDUAL SOLAR ROTATIONS

Fig. 3.2. A recent example of the sunspot "butterfly" diagram, which displays the time and latitude variations of the appearance of new sunspots during successive solar cycles. As Carrington discovered, the first spots in a new cycle appear around latitude 30 in both solar hemispheres. As the cycle advances, new spots appear closer and closer to the solar equator.

Hale continued to measure the strength and magnetic polarity of sunspots for several years after his groundbreaking discovery. In 1919 he summarized his findings in a series of rules that the sunspots seem to follow.

First, sunspots usually appear on a line approximately east-west, in pairs of opposite polarity. The sunspot farther to the west (the direction of solar rotation) was named the "leader"; the other was called the "follower." The follower usually emerges slightly poleward of the leader.

Next, in any cycle of eleven years, most of the leaders in the Sun's northern hemisphere have the same magnetic polarity, which is opposite to that of the leaders in the southern hemisphere. In the next cycle, the first new spots appear at high solar latitudes and all the leaders have now reversed their polarity. A complete magnetic cycle therefore requires twenty-two years.

Taken together with the discoveries of Schwabe and Carrington, Hale's "laws" implied that the Sun has an internal mechanism for generating magnetic fields in a highly systematic cycle, whose origin theorists would struggle for decades to explain. In truth, they are still struggling.

Some enterprising folks haven't waited for a scientific explanation, however. They've sought signs of the cycle's influence on any and all human affairs. As we shall see, the cycle is seriously thought to affect our climate, if not our weather. But more speculatively, the quality of French wines, the price of wheat, the length of women's dresses, the business cycle, and the alternation of political parties have all been attributed to the influence of the number of sunspots. It doesn't take a doctorate in physics to use (or misuse) the products of science.

But I digress. Let's return to the serious search for an understanding of the cycle.

MODELS OF THE SOLAR CYCLE

We saw in chapter 2 how theorists such as Edward Bullard, Walter Elsasser, and Eugene Parker developed the outlines of a theory for the maintenance of the Earth's magnetic field. They proposed that the field is generated by a self-excited dynamo in the Earth's liquid metal core. The rotation of the electrically conducting core through an existing poloidal field would generate a toroidal field. This process is now called the omega effect. Parker then introduced the alpha effect, in which cyclonic convection cells push up segments of the toroidal field and twist them into the meridional plane. Parker showed how many small loops might coalesce into the nearly steady dipole field we detect at the Earth's surface.

In the same 1955 paper in which Parker introduced cyclonic convection, he also began to explore a dynamo for the Sun. Unlike the Earth the Sun has a relatively short magnetic cycle, which he had to explain. He proposed to locate a self-excited dynamo in the Sun's convection zone, the thick layer that extends from the surface (the photosphere) to a depth of about a third of a solar radius. The internal properties of the zone were uncertain at the time, but convection cells (called granules) were known to rise and cover the entire surface (note 3.3).

To make progress, Parker had to guess how the interior of the convection zone rotates. At the surface of the Sun, the equator rotates in about twenty-seven days, while latitude 70 degrees requires thirty-two days. (Christoph Scheiner, a Jesuit priest and a contemporary of Galileo, discovered this behavior around 1630 by tracking sunspots.) In 1955 this "differential rotation in latitude" was generally interpreted to mean that the angular speed of rotation is constant on cylinders centered on the Sun's axis and increases with distance from the axis. Parker left open the question of just how the rotation varies but adopted the consensus view as an illustration.

He then demonstrated mathematically that a depth-variation of rotation, when coupled to cyclonic convection cells, would generate a migratory "dynamo wave" in each hemisphere. One can picture each wave as a series of twisted flux ropes that circle the Sun on parallels of latitude. Each rope consists of a strong toroidal field and a number of poloidal loops that add the twist. The assumed gradient of rotation speeds causes the tops of the loops to rotate faster than their bottoms; they tip over, shear, and stretch in the direction of the toroidal field. On one side of a rope, the stretched loops add to the existing toroidal field; on the other side, they cancel it. This effect causes the toroidal field in the ropes to migrate in latitude.

A rope starts as a weak torus at high latitude and is amplified (by the omega effect) as it approaches the equator. Simultaneously, convection cells regenerate the loops by the alpha process. The ropes in each hemisphere meet at the equator at the end of a cycle, and because they have opposite polarity, they cancel. At this moment the next pair of ropes is supposed to launch at high latitudes.

This was hardly a complete model of the Sun's magnetic cycle. But Parker pointed out the similarity of the migrating dynamo wave he had discovered to the latitude migration of the sunspot belts during a cycle.

Next he had to explain how such ropes could produce sunspots. He had another bright idea in 1955. He recognized that the magnetic pressure inside a horizontal flux rope would squeeze out some plasma. With lower plasma density inside, the tube would be buoyant, like a balloon under water, and would tend to rise.

Parker showed that this situation is unstable. A weak point somewhere on the horizontal tube would develop a buoyant kink, in the shape of a Greek omega (Ω), which would float up to the surface. When it broke through the surface, its two legs would remain tied to its parent tube. Therefore, the magnetic field in one leg would point out of the surface; the field in the other leg would point inward.

The stage was now set for the formation of a bipolar pair of sunspots. First, though, Parker needed an additional mechanism to concentrate the field lines in each leg of the kink and so create the strong fields (up to 3 kilogauss) that Hale had observed. He recalled a key paper that Ludwig Biermann, a German astrophysicist, had published in 1939.

Biermann was an eminent scientist who had done important research on convection in stars and had a side interest in the physics of comets. In later life he rose to direct the prestigious Max Planck Institute for Astrophysics in Munich, Germany. A shy, formal man with a stammer, he had the gift of asking incisive questions. He will enter our story on several occasions.

Biermann had asked himself a very simple question: "Why are sunspots darker than their surroundings?" He knew that a strong magnetic field could inhibit the circular motions of convection cells near the surface (note 3.4). Energy supply to the spot would therefore be less efficient and the spot would cool and contract. The contraction would further concentrate the sunspot's field lines, leading to further cooling, until a balance of magnetic and plasma pressure was reached.

Parker invoked Biermann's cooling mechanism to complete his scenario for the formation of a pair of sunspots with strong magnetic fields. By the late 1950s Parker had sketched a rudimentary theory for the solar cycle and the formation of sunspots, but new observations at Mount Wilson Observatory would soon require a thorough revision of his basic ideas.

Babcock's Observations and His Model

Hale's reliance on photography limited him to observing only the strongest magnetic fields on the Sun. Astronomers had to wait until the 1950s for the tools that would allow them to investigate the weaker fields. In that decade scientists in the Soviet Union, Germany, and the United States broke new ground.

Among the leaders were the father and son team of Harold and Horace Babcock, astronomers at the Mount Wilson Observatory. Horace, the son, was a talented experimenter, familiar with the latest in technology. In the late 1950s he built an electronic instrument (a magnetometer) that allowed him to detect fields of only a few hundred gauss, ten times weaker than sunspot fields.

He discovered a new world of complexity. Magnetic fields are not confined to the dark sunspots but spread over much of the surface. Around each pair of sunspots is a broad area of enhanced field strength. These "bipolar magnetic regions" (BMRs) survive long after the spots within them have disappeared. They expand as they age and seem to be dragged out into long streaks by differential rotation, the latitude variation of rotation speed. During the solar cycle, most BMRs drift toward the equator, while some at higher latitudes migrate rapidly toward the nearest pole and reverse the polarity of its weak field.

Horace Babcock thought long and hard about the many facets of the sunspot cycle. How could he explain what he and Hale had discovered? He was not primarily a theorist, but he had good physical intuition and could build on earlier research by others. In 1961 he proposed a scenario that could account for many of the observations, but without much quantitative detail.

In his picture, the cycle begins with the Sun having a simple poloidal field, similar to the Earth's. Some of the field lines are supposedly submerged in the plasma just below the surface. Next Babcock proposed that differential rotation drags these shallow poloidal field lines into a toroidal field with slightly sloping east-west components (see fig. 1.4 and accompanying text). As we shall see, the slope of the toroidal field lines is a critical feature in Babcock's scheme.

With successive rotations the field lines wrap around the Sun many times, thereby building up the intensity of the submerged toroidal field. Differential rotation also twists the subsurface field lines and forms "ropes" of magnetic flux. Following Parker's scenario, Babcock supposes that at some critical field strength a kink or loop in a rope rises buoyantly and breaks through the surface. The loop's ends are now anchored in a BMR, which consists of a pair of sunspots and the surrounding weaker fields. Because the submerged ropes slope poleward in each hemisphere (see fig. 1.4 again), the following parts of the BMRs emerge poleward of the leading parts. This difference becomes important in what follows.

As the cycle progresses, the critical field strength for buoyancy is reached at lower and lower latitudes, so that new BMRs appear closer to the equator. The leading part of one region connects to the following part of another to form a coronal flux loop that escapes to space. Most of the flux generated in a cycle is supposed to disappear this way.

Babcock invokes a meridional flow to drive the following parts of BMRs toward the nearest pole. Their magnetic polarity is opposite to the pole, and because they were born closer to the pole, they arrive there long before any leaders do. When they arrive, they reverse the magnetic polarity of the pole. A reversed dipolar field begins to form in the corona. In the declining phase of the cycle, the leading parts in both hemispheres meet at the equator, and having opposite polarities, they form loops that span the equator. These new loops lay essentially north-south and point in the direction of the new dipolar field. Babcock requires them to sink at the equator and contribute to the formation of the new dipolar field.

Babcock's model was novel in several important respects. Unlike Parker's model, it relies entirely on processes at or just under the surface in a thin layer. The transport of flux over the surface was critical in reversing the polar fields. Although he used the familiar omega process of differential rotation to amplify the toroidal field, he added a plausible slope to the field lines. Most important, he suggested a new alpha mechanism (the submerging transequatorial loops) to generate the new dipolar field.

His model did rely on some uncertain effects, however. He had to postulate the existence of poleward and equatorward meridional flows to account for the drift of the BMRs. (A weak polar flow was observed twenty years later.) Also, he had to assume the equatorward flow would submerge flux loops despite their buoyancy. In spite of its uncertainties, his model has had a lasting effect on the subject. It was soon extended and improved by a colleague of Babcock's, Robert Leighton.

Supergranules and Surface Flux Transport

In the 1960s, Robert Leighton was generally regarded as one of the most inspiring physics teachers in Caltech's physics department. He began his career as an experimental cosmic ray physicist but gradually moved into solar physics and later into infrared and radio astronomy. In 1962 he modified some solar equipment at the Mount Wilson Observatory and with the help of several graduate students made three stunning discoveries.

First, they found that patches all over the surface of the Sun oscillate vertically with a period of five minutes. Second, they discovered a random pattern of convection cells that covered the surface of the Sun. These "supergranules" were up to twenty times larger than the well-known "granules" but were too faint and too slow to be easily detected. A supergranule emerges, expands horizontally, and merges with its neighbors within about a day. Inside each of these cells, the plasma rises at the center, flows to the edges, and sinks.

Leighton and his troupe made a third discovery, which involved the magnetic field at the Sun's surface. Any field lines that thread vertically through a supergranule cell are pushed to its boundary. As new cells emerge and decay, a field line can be shuffled long distances over the surface in a "random walk" (note 3.5). In this process, field lines tend to collect in the supergranule boundaries, where they join an open network (like a fisherman's net) that covers the Sun.

Leighton realized that the combined effect of supergranules was to disperse magnetic flux from BMRs, where the flux was concentrated, to field-free surrounding areas. Here was an observable process that could account for the spread and migration of a magnetic field over the solar surface.

He also recognized the importance of the slope of the subsurface toroidal field that Babcock had introduced. The slope implies that each erupting flux loop has both a strictly east-west (toroidal) component and a smaller north-south (poloidal) component. The poloidal components of many loops could be assembled by appropriate means to reverse the dipole field of the Sun.

Leighton constructed a semi-empirical model based on Babcock's, in which all motions were specified in advance and the fields were averaged in longitude and in depth. With a careful choice of nineteen parameters, Leighton could reproduce many of the observed characteristics of the cycle. Critics of the model noticed, however, that cyclonic convection is virtually ignored, a distinct difference from the Parker dynamo. This omission was considered a serious flaw, because convection is actually observed in the very surface layers described by Leighton.

As we described in chapter 2, Max Steenbeck, Fritz Krause, and Karl-Heinz Rädler developed a theory for the geomagnetic field, called mean-field magnetohydrodynamics. Around 1960 they applied it to the Sun. It was a mathematical representation of a self-excited dynamo operating in the deep convection zone. Although all the relevant motions were specified in advance, cyclonic convection did couple with differential rotation in latitude and depth to produce a solar cycle. Such mean-field models rapidly displaced the Babcock-Leighton model.

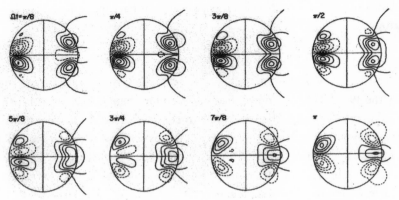

Fig. 3.3. These diagrams show the evolution of magnetic fields during an eleven-year solar cycle of an alpha-omega dynamo. The calculated field has been averaged in longitude. Each panel shows a cross-section of the Sun; the vertical line is the axis of rotation. On the left side are contours of equal toroidal field strength. The actual field lines lie along circles of latitude. On the right side are poloidal field lines, in meridional planes. Solid lines indicate positive toroidal field and clockwise poloidal field lines; dashed curves indicate the opposite. Notice how all fields appear at high latitude and march toward the equator during the cycle, in accord with observations of sunspots.

Figure 3.3 is a typical example of an alpha-omega dynamo model, calculated by Michael Stix in 1976 with the mean-field formalism. It shows the evolution of the fields at eight stages through an eleven-year cycle.

Notice how each type of field migrates toward the equator as the cycle progresses, in accord with Carrington's observations of sunspot births. This is an example of Parker's dynamo wave. Newly generated flux appears poleward of the existing flux, grows in magnitude as it migrates equatorward, and gradually displaces this older flux. This model does not reproduce the observed polarity reversal of the poles very well but is otherwise impressive.

To build a model like this, Stix had to guess how the rotation speed varies in the convection zone. He made the reasonable assumption that the rotation rate increases with increasing distance from the Sun's axis of rotation. That would immediately explain why, at the surface, the equator rotates faster than the higher latitudes. But when he plugged this assumption into the mean-field theory, it predicted that spots would be born at increasing latitudes as the cycle advanced, which is opposite to the observations. Only a rotation rate that decreases with distance from the axis led to acceptable results, as shown in the figure, but such a rotation rate was hard to justify on physical grounds.

The Whole Ball of Wax

By the 1980s computer power had developed to the point where a theorist could try to simulate the solar cycle in complete detail, starting with the basic physics of convection. Several courageous souls have spent a good part of their careers since then following this approach. Peter Gilman is one such, working at the High Altitude Observatory in Boulder, Colorado. Gary Glatzmaier, whom we met earlier, is another.

Each researcher independently constructed a 3-D time-varying computer model of a rotating, convecting, magnetized sphere of compressible plasma. As you can imagine, the task required thousands of hours of effort. The researchers made no assumptions on how rotation varies in depth or latitude. Instead the differential rotation would emerge spontaneously from the interaction of the convection with the Sun's twenty-seven-day rotation.

When each simulation was run on a computer, the model predicted the angular rotation speeds actually observed at the surface. That was a great success. But each model also predicted that new flux (corresponding to sunspots) would appear at increasing latitudes as the cycle progressed, in conflict with Carrington's results. The reason was that the elaborate models predicted that rotation speeds decrease with increasing depth. As Stix had learned earlier, this was bound to yield the undesired result. What could they change in their models?

The issue was settled with fresh observations, as a radically new technique called helioseismology was discovered.

The Vibrating Sun

In 1962, Robert Leighton and two of his students at Caltech discovered that patches all over the surface of the Sun oscillate with a period of about five minutes. At first it appeared that each patch was oscillating independently of all the others, in some kind of random process. But Roger Ulrich at UCLA soon offered a different explanation.

He pointed out that convection in the thick envelope of the Sun is noisy; it generates sound waves of all wavelengths. These waves bounce around between the surface and the bottom of the convection zone and interfere with each other. Some waves reinforce each other; others cancel. The result is a complicated pattern of standing waves that fills the convection zone. At the surface, we see what appears to be a jumble of oscillating patches but is in fact a perfectly ordered pattern.

A few years later Franz-Ludwig Deubner, a German astronomer, made the critical observations that confirmed Ulrich's theory, the implications of which were huge. According to the theory, the observed pattern ultimately depends on the way temperature and rotation speeds vary in the Sun's interior. Therefore, by observing the pattern at the surface for months or years and applying the theory, observers could determine the temperature and rotation throughout the convection zone. A new science, helioseismology, was born. After two decades of intensive observations, the complete internal rotation and temperature layering of the solar interior was determined.

Theorists everywhere were stunned by the empirical results. Rotation speeds are not constant on cylinders that are centered on the Sun's axis, as previously thought. Instead, the angular rotation rates are constant along a radius from the center of the Sun and decrease with latitude. In other words, if you bored a vertical hole from the top to the bottom of the convection zone, at a chosen latitude, all points along the hole would rotate at the same angular speed around the axis of the Sun. That speed would depend on the latitude you chose, with the highest speed at the equator.

Below the convection zone lies a "radiative zone," a spherical shell that has a thickness of about three-tenths of the solar radius. Observers were surprised to learn that the zone rotates like a solid body, with a rate equal to the surface rate at latitude 30 degrees.

Now for the critical part: the bottom of the convection zone and the top of the radiative zone meet in a thin interface, the tachocline. The rotation rates are different on opposite sides of the joint, so the two zones must be rubbing against each other there, and as we shall see, that conclusion has important consequences for the generation of toroidal magnetic fields.

These surprising results challenged theorists like Juri Toomre and his colleagues at the University of Colorado to understand the physics that underlies them. Toomre decided to postpone an attack on the solar cycle problem until he and his crew better understood the origin of the Sun's rotation.

Toomre had been studying turbulent convection in a rotating star for at least two decades. His method is numerical simulation. He chooses an imaginary container, like a box or a sphere that is filled with plasma. Gravity acts on the plasma. He allows heat to flow in at one boundary and out the other, and using classical fluid dynamical laws, he calculates the velocity at every point as a function of time. He makes as few assumptions on the physics as he can, which means he needs an enormous computer code. The number of spatial points his computer can handle in a reasonable time limits his ability to follow the evolution of turbulent eddies.

He began with relatively simple situations, like convection in a rectangular box. Then he added rotation and watched to see how convective flows are affected by Coriolis forces. Now he and his students are using the fastest parallel processing computers available to calculate rotation rates throughout the Sun from basic physical principles. Their huge computer codes incorporate much of the known physics of convection and turbulence.

Mark Miesch, a mathematical physicist at the University of Cambridge joined Toomre, Glatzmaier, and colleagues in 2000 to carry out a monumental simulation of solar convection. This project had all the bells and whistles, including perhaps the most highly resolved turbulent cells. Their results revealed a zoo of complicated flows, including a network of transient downflows in the upper convection zone and strong vertical downflows in narrow lanes. They hoped to reproduce the rotation patterns derived from helioseismology, but the match between their models and the helioseismic data was still not perfect.

Nevertheless, one might think that the next obvious step would be to include magnetic fields in a numerical simulation like this and see if one could reproduce the main features of the sunspot cycle. That strategy has run into a number of roadblocks, first pointed out by such researchers as George Fisher, Arnab Rai Choudhuri, Sydney D'Silva, and Edward DeLuca.

They explored a critical step in the dynamo process, the rise of toroidal flux tubes from the deep convection zone to the surface, to form omega loops. They discovered that only a tube with a magnetic field strength of about 100,000 gauss could emerge at the surface in the latitudes where sunspots are observed.

Any tube weaker than that emerges too close to the pole. Also any tube weaker than that gets chewed up by the convection. But tubes with such strong field strength are exceedingly buoyant and would float up in only a few months. That is far from the eleven- or twenty-two-year period of the solar cycle. The researchers were stuck between a rock and a hard place.

That was the bad news. The good news was the new pattern of rotation that helioseismology revealed. It confirmed that the bulk of the convection zone is unlikely to generate toroidal fields, because its pattern of rotation would not stretch and wind magnetic field lines. The only place where such effects could occur is across the tachocline, the thin boundary layer between the convection and radiative zones.

Moreover, theorists realized that the radiative zone would be an ideal place to store 100,000-gauss flux tubes almost indefinitely. That's because the zone has a very gradual temperature gradient that suppresses the buoyancy of a flux tube (note 3.6). So the problem was reduced to generating strong flux tubes in the tachocline, storing them deeper down in the radiative zone, and allowing them to leak very slowly or diffuse to the surface.

Toward a Model of the Solar Cycle

Eugene Parker rose to the challenge once again. He realized in 1993 that a critical factor in the dynamo is the "diffusivity," which controls the speed at which flux can leak through plasma. Estimates of the diffusivity suggest that it is very small in the stable radiative zone and a hundred times larger in the turbulent convection zone. He showed how that condition could lead to a separation of the two critical processes in the dynamo.

The omega process of generating toroidal field could occur just below the tachocline, and the alpha process of generating poloidal field could occur just above it. The jump in the value of the diffusivity would control the leakage of magnetic flux through the tachocline. Such an "interface dynamo," he suggested, might be possible at the tachocline.

Keith MacGregor and Paul Charbonneau at the High Altitude Observatory in Boulder tested the idea with numerical simulations in 1997. They calculated a variety of interface dynamos, using the rotation speeds determined from helioseismology. On the whole their results did support Parker's basic scheme.

That's not the end of the story, though. In 2005, Mausumi Dikpati, Peter Gilman, and Keith MacGregor, all from the High Altitude Observatory, pointed out a serious flaw in an interface model. Without an additional meridional flow along the tachocline, they claim, an interface model will not generate the cyclical migration of sunspots we see at the surface.

Meridional flows along the tachocline have not been observed as yet, and indeed, if they exist, they must be extremely slow and difficult to detect. At the surface, a steady poleward flow of only 10 to 20 m/s has been measured, about one hundred times smaller than is common in convection cells. Nevertheless, these recent conclusions have forced the Boulder researchers to reconsider an old model, the Babcock-Leighton scenario, in which meridional flows are essential.

In that scenario, a meridional flow was postulated to carry poloidal flux to the pole and reverse its polarity. In the latest interface models, that poleward flow would sink deep down to the tachocline and return along a meridian to the equator. As poloidal flux is carried along the tachocline, it is sheared by the latitudinal gradient of rotation. Flux ropes rise back to the surface and all proceeds as before.

Most recently Dikpati and Gilman have demonstrated some convincing solar cycle models, which are blends of the interface and flux transport models. They are so confident in their work that they have offered predictions for the next solar cycle.

That is not to say the solar cycle problem is solved or that other ideas are not flourishing. Indeed, the ultimate demonstration that we understand the physics of the solar cycle could require a full-scale numerical simulation, such as Gary Glatzmaier performed for the Earth's magnetic field. In principle such a simulation would reproduce all the observed features, including the century-long cycles and the occasional disappearances of the cycle (note 3.7). A comprehensive theory could also predict the cycle's future variations. That ambitious goal is still some distance away.

SUNSPOTS REVISITED

Over the past forty years, as a hardy band of theorists has labored to understand the machinery of the solar cycle, solar observers have been opening a vast new territory. Advances in telescope technology, digital imaging, and electronics have enabled them to measure vector magnetic fields and surface plasma flows at higher resolution and with more precision than Babcock and Leighton could ever dream of. Moreover, the techniques of helioseismology have been extended to localized structures like sunspots and active regions. Now it is possible to explore the thermal and magnetic properties of sunspots beneath the visible surface.

Some of the most bizarre and fascinating results have emerged from recent high-resolution observations of sunspots. They reveal a variety of waves, oscillations, and flows that challenge the best minds. We'll spend the rest of this chapter exploring them. Let's begin with some basic background.

A Sunspot Primer

Around 1630, Galileo Galilei and his contemporary Christoph Scheiner argued bitterly over who had first discovered sunspots and what they actually were. Scheiner defended his view that they were merely dark clouds hovering over an unblemished Sun. The idea that a celestial body could have warts was unpalatable to him, as a Jesuit priest trained in Ptolemy's orthodoxies. Galileo insisted that the spots changed shape in a way that proved they were objects on the Sun's surface. Galileo won the argument, but Scheiner won a reputation for the beautiful drawings of sunspots he made and for his discovery that the Sun's rotational axis is tipped with respect to the ecliptic. In his drawings of a fully grown sunspot, we see a dark central umbra surrounded by a feathery penumbra.

We fast-forward to the era of George Ellery Hale and the Mount Wilson Observatory. Hale and his colleagues established that sunspot sizes range from tiny "pores" only 2,000 km in diameter to giants 50,000 km or more. A large spot has umbral fields as strong as 3,000 gauss and can survive as a recognizable object for several months. Pores have no penumbras, their centers are not as dark as large spots, and their fields are also generally weaker. They live for only about a day.

In some of the sharpest photographs of sunspots made at Mount Wilson, the photospheric penumbra is resolved into narrow filaments that radiate from the umbra like the spokes of a wheel. In a few cases the filaments are curved and seem to whirl around the umbra like a vortex. As we learned earlier, the vortices Hale saw more clearly in the sunspot chromosphere led him to look for magnetic fields.

In 1909 British astronomer John Evershed was observing at the Kodaika-nal Solar Observatory in southern India. On one auspicious morning in January, he discovered what he later called "radial movement in sunspots" (*Observatory* 32, 291, 1909). He measured a radial penumbral outflow at about a kilometer per second and radial inflows in the umbral chromosphere at a few kilometers per second. As we shall see, the details of the flows are complicated and the full paths of this circulation have still not been traced, as far as I know.

Hale and his colleague Seth Nicholson measured the gross magnetic structure of round sunspots. They reported in 1938 that the field is symmetrical around the umbra's axis; is vertical on the axis; and emerges at greater angles to the vertical at increasing distances from the axis. The field is nearly horizontal at the outer edge of the penumbra. So, if you like, the field lines resemble the branches of a willow tree, with the lower boughs drooping on the ground.

Observers have measured the brightness of sunspots at different positions on the solar disk in order to determine their temperatures as a function of depth and radius. A typical umbra's surface temperature is about 3,900 kelvin (K), almost 2,000 degrees cooler than the surrounding photosphere. That, of course, is why it is dark. The penumbra is also cooler, but not as much; it has a typical temperature of about 5,000 K. Knowing the vertical profile of temperatures and the shape of the magnetic fields, a theorist can construct a simple static model of a sunspot, if he assumes that gas and field are in pressure equilibrium everywhere.

Recent observations show that a sunspot grows by accumulating the small pores around it. After a few days, it reaches its maximum size and begins to decay. Observers are divided on the actual process of decay. Some favor a submergence of flux; others emphasize a gradual nibbling away of the spot's flux by granules at its edges.

Sunspots under the Microscope

In figure 3.4 we see one of the sharpest digital images of a sunspot ever captured in white light. (Galileo and Scheiner would be astounded, I'm sure.) It was obtained in 2005 by German scientist Friedrich Woeger with the help of adaptive optics, a marvelous invention now utilized at many telescopes to

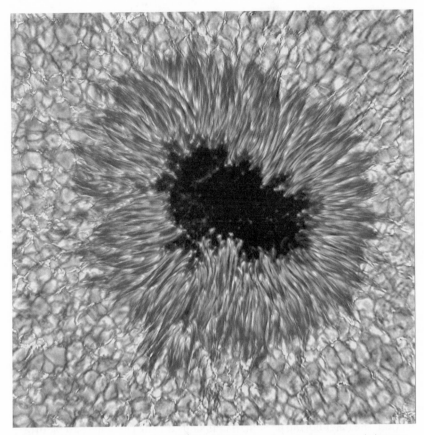

Fig. 3.4. Here we see one of the sharpest white-light images of a sunspot ever recorded. It was obtained at the Sacramento Peak Observatory with the aid of adaptive optics that remove the blur introduced by the turbulent atmosphere. The penumbral filaments are less than 100 km wide.

remove the blurring effects of the Earth's atmosphere. We are looking through the atmosphere of the sunspot, down to the level of the photosphere. The umbra of this sunspot is about three times the diameter of Earth. The penumbral filaments are revealed with stunning clarity. The narrowest ones are at the limit of the telescope's resolution, about 80 km in width, or a fortieth of the umbra's diameter.

White-light pictures like these are very helpful, but to nail down the physical properties of a sunspot, an observer needs its spectrum. Some spectral lines are sensitive to temperature, others to magnetic field, still others to motions. Moreover, each line originates within a relatively narrow range of depths in the sunspot atmosphere. By choosing several spectral lines with some care, an observer can probe the entire volume of a sunspot atmosphere (note 3.8). And with the development of adaptive optics, he or she can obtain spectra with almost the same high spatial resolution she can achieve in white-light images.

From such data, an observer can construct a three-dimensional empirical model of a sunspot. Several groups have been following this approach for over a decade. They have opened a Pandora's box of problems that are challenging the best theorists. The penumbra is the most difficult puzzle. Only recently has a comprehensive picture of the penumbra begun to emerge.

Alan Title and his team from the Lockheed Palo Alto Laboratory revolutionized our ideas about the penumbra in 1987. They were observing sunspots with Gören Scharmer, at the Swedish Solar Observatory in La Palma (note 3.9). The evacuated telescope there was delivering solar images of unprecedented sharpness, even without adaptive optics. Title was using a clever tunable filter he had invented and making movies of the flows in penumbral filaments.

He discovered something remarkable: penumbral filaments don't all lie in the same horizontal plane, as they appear to do in figure 3.3. Instead, alternate bright and dark filaments have different inclinations with respect to the vertical direction. The bright filaments slope upward in the outward direction; the neighboring dark filaments slope down. Title offered a picture of two combs with interlocking teeth as a visual guide to the structure.

In the past decade, Title's comb structure has been extended and revised by many investigators, including Bruce Lites at the High Altitude Observatory; Jürgen Staude at the Potsdam Solar Observatory; Sami Solanki at the Max Planck Institute for Aeronomy in Lindau, Germany; and Thomas Rimmele at the National Solar Observatory. Their work has given us the following tentative portrait of the penumbra.

Penumbral filaments are magnetic tubes that emerge near the umbra and extend outward toward the edge of the sunspot. Dark and bright filaments are members of two distinct groups that interlock (Title's comb) but don't interact during the life of the sunspot. The dark filaments are everywhere more inclined to the vertical than the bright filaments. For example, at the

umbra-penumbra boundary the bright filaments emerge at 30 degrees inclination, the dark ones at about 60 degrees. Both groups incline farther from
the vertical as the sunspot's outer edge is approached. Exactly at the edge,
bright filaments incline at 70 degrees and dark filaments are horizontal. Images captured by TRACE show that the bright filaments extend well beyond
the sunspot. Where they end is uncertain.

The Evershed outflow is concentrated in unusually long dark filaments.
Thomas Rimmele has observed such filaments extending as much as 10,000 km
beyond the edge of the penumbra and diving into the photosphere there. As
the filament crosses the edge, it reverses from being darker than the photosphere to brighter.

As if all this were not complicated enough, the dark filaments arch over
the photosphere for much of their length, forming a kind of magnetic canopy.
In 2001, Bruce Lites and friends were able to measure the field strength in
different layers of the penumbra. They discovered that the canopy has strong
fields (800 gauss) that cover relatively weak horizontal fields (300 gauss) nearer
the surface.

Since 2002, observations at the Swedish Solar Vacuum Tower, and the
Dutch Open Telescope in La Palma, have resolved the internal structure of
a filament. Bright filaments consist of a dark core and a bright sheath that
seem to flow as a coherent unit. These latest results also beg for a physical
interpretation.

Much hard thought has been expended in trying to understand the origin
of the Evershed flows. In 1968, Friedrich Meyer and Hermann Schmidt, two
associates of Ludwig Biermann, proposed that both the outward flow in the
photosphere and the inward flow in the chromosphere are siphon flows, driven
by differences of pressure at the ends of a filament. Their model, with later refinements, has held up remarkably well.

However, John Thomas at the University of Rochester and Nigel Weiss at
the University of Cambridge realized that the outer end of a dark filament
(its footpoint) would be buoyant in the photosphere and would require some
force to anchor it there. They proposed that the filament's field lines are held
down by the vigorous downflows observed between granules. They demonstrated the feasibility of this pumping mechanism with numerical simulations, but the issue is still controversial.

Sunspot Oscillations and Internal Structure

We learned earlier how Robert Leighton and his students discovered that the Sun's surface is entirely covered with patches that oscillate vertically with a period around five minutes. So it may not be surprising to learn that sunspot umbras also oscillate with a five-minute period, although with lower amplitude. But Ronald Giovanelli, an Australian astrophysicist, discovered in 1972 that umbras also oscillate at three minutes, in both the photosphere and chromosphere. And that poses a problem. A period half of five minutes might be expected, as the first harmonic of five minutes, but three minutes is puzzling.

At least it was to John Thomas and Alan Nye at the University of Rochester and their colleague Lawrence Cram at the National Solar Observatory. In 1982 they recognized that this unexpected pair of periods offered clues to the internal structure of a sunspot. The five-minute oscillation of an umbra required relatively little explanation. The photosphere forces an umbra to oscillate at this period. In effect the photosphere periodically squeezes the bundle of field lines rising out of the umbra and drives the oscillation.

The steady three-minute umbral oscillation, on the other hand, is likely to be a resonant response. A resonance like this implies the existence of a volume (misnamed a "cavity") in which waves are trapped and in which a standing wave can be maintained. The umbra behaves in some sense like a big drum with a definite tone.

Thomas and friends realized that the three-minute period is a clue to the dimensions of the cavity within a sunspot. Therefore one might be able to use the oscillations as a probe of sunspot structure. In a sense this technique would be an extension of helioseismology to the structure of sunspots.

Thomas and colleagues built on the work of previous investigators, especially Jürgen Staude and Yevgeni Zhugdha at the Potsdam Solar Observatory. In 1987 they developed a two-cavity model to account for the three-minute oscillations. The first step was to identify the particular type of wave that is trapped in these cavities. They decided the most likely candidate was a "magnetosonic" wave, a hybrid of a sound wave and an Alfvén wave.

Let's take a moment to understand these waves. A pure sound wave (a single tone) propagates as a periodic variation of plasma pressure. In uniform plasma it propagates in all directions at a speed determined by the plasma temperature. In contrast, a pure Alfvén wave propagates only along field lines as a periodic variation of magnetic tension. Its speed is determined by the magnetic field strength and the plasma density. Stronger fields and lower densities lead to faster Alfvén speeds.

In a flux tube filled with plasma, a magnetosonic wave can propagate as a "slow mode" along the field lines at the speed of sound. Or it can advance as a "fast mode" at any angle to the field direction and at a speed that is a combination of sound speed and Alfvén speed.

Thomas and friends proposed that the two resonant cavities in a sunspot umbra are shaped by the height-variations of sound speed and Alfvén speeds. For simplicity they assumed that the umbral field is uniform and vertical and that the plasma density decreases with height exponentially. Earlier studies had shown that the umbral temperature falls to a minimum of about 3,900 K at about 600 km above the solar surface and increases above and below.

These researchers demonstrated that fast modes are trapped in a cavity in the deep photosphere. They are reflected downward by the rapid increase of Alfvén speed with increasing height, and reflected up by the rapid increase of sound speed with increasing depth. In the chromospheric cavity, slow modes are reflected down by the rapid rise of sound speed with height, and reflected up by the temperature minimum.

The two cavities are separated in height by a no-man's-land in which both modes fade out. However, a small amount of wave energy can leak from each cavity, reach the other cavity, and change modes. This type of conversion produces a bewildering variety of interactions that seriously confused observers.

Thomas Bogdan at the High Altitude Observatory is one of a number of theorists who has taken on the task of sorting out this tangle of oscillations. He uses numerical simulations of increasingly realistic magnetic models of a sunspot and explores the variety of mode conversions that are possible. The details are fascinating and have attracted numerous theorists to the subject, but we will have to leave them here.

In 1987, Douglas Braun, Thomas Duvall, and Barry LaBonte discovered that a sunspot absorbs about half of the power it receives from the five-minute photospheric oscillations. Their result pointed to the source of the power in the three-minute oscillations but raised the question of the conversion and dissipation of this power. It now appears that some of this absorbed energy escapes from the umbra as waves that propagate into the corona.

Indeed, three-minute oscillations have been detected in the high chromosphere and corona by instruments aboard SOHO and TRACE. They grow stronger in amplitude with increasing height, up to the level above a sunspot where the temperature reaches 100,000 to 200,000 degrees and then fade at higher temperatures. By correlating their phases at different heights, observers have determined that they are slow modes propagating along the extended sunspot magnetic field. The amount of energy they carry, though, is far too small to account for the cooling of the sunspot.

We can't leave the subject of sunspot waves without mentioning the running penumbral waves. They were discovered in 1972 by Harold Zirin and Alan Stein at Caltech, and independently by Ronald Giovanelli. They saw a bright wave propagating radially outward in the penumbral chromosphere at about 10 km/s, preceded by an "umbral flash."

A few years earlier, Jacques Beckers and Paul Tallant at Sacramento Peak Observatory had discovered umbral flashes. They are pulses of light that occur irregularly at several umbral locations at intervals of around 125 seconds; they last for about 50 seconds and propagate toward the penumbra at 40 km/s.

In 2003 observers at the Swedish Solar Vacuum Tower and the Dutch Open Telescope at La Palma confirmed that flashes in the umbral chromosphere seem to initiate penumbral running waves. Moreover, the motions in the flash suggest that flashes are upward shocks.

In their numerical simulations, Thomas Bogdan and associates have seen running penumbral waves that are initiated by sharp pulses in the umbra. They identify the waves as slow modes that can be modified by interactions with fast modes that impinge from above. But the physical connections between the flashes and the running waves are still uncertain.

Sunspots Undercover

What does a sunspot look like underneath the photosphere? For lack of any observations, modelers made the simplest assumption. They pictured a monolithic column of magnetic flux, standing upright in a field-free convection zone. The diameter of the column would have to shrink at increasing depths, so that its magnetic pressure could balance the increasing external plasma pressure.

Such a model has trouble supplying enough energy to a sunspot. Heat cannot easily pass sideways into the column by conduction, nor can convection cells squeeze between the packed field lines.

So in 1975 Eugene Parker proposed that below the surface the umbral field divides into many separate tubes with ample spaces among them. The sunspot would look like a Portuguese man-of-war, with a tangle of drooping tentacles. His model would allow convection cells to supply heat more easily to the upper layers, where it is needed to maintain the sunspot.

Such "cluster" models were revised and debated in the following years. Karl Jahn and Hermann Schmidt proposed a radically different model in the 1990s. They suggested that a thick penumbra delivers at least some heat to a sunspot. In their models, nearly horizontal filaments rise and sink in continuous slow motion, carrying heat to the surface and radiating it to space, but the flows involved in such convection were not spelled out.

Until recently there was no way to choose among models. But new techniques of helioseismology have been developed in the past decade that allow an observer to look underneath a sunspot or an active region. These advances date back to 1987, when Braun, Duvall, and LaBonte invented "local helioseismology" during their study of the absorption of waves by sunspots. Later, Duvall invented another clever method, time-distance tomography. These tools have opened a window beneath the surface of the photosphere. In fact, it is now possible to image sunspots on the invisible side of the Sun with the five-minute oscillations. Be aware, though, that even these sophisticated methods cannot detect magnetic fields except by their effects on the oscillations.

In 2004, Alexei Kosovichev at Stanford University acquired observations of the oscillations in and around a sunspot from an instrument aboard SOHO. Then he analyzed the data using time-distance tomography. He was able to reconstruct the temperatures and flow speeds down to a depth of 24,000 km. He found a thin, cool upper layer (the umbra) and a hotter layer deeper down. Converging downflows are seen in the upper layer and diverging upflows in the lower layer. Kosovichev concluded that his results are consistent with Parker's model because it predicts shallow converging flows. However, some other models envision diverging upflows in the deep layers. So although the issue is still open, we can expect it to be resolved as observations improve.

Sunspots and Sunshine

Before we leave the subject of sunspots, we should mention their effect on the energy we receive from the Sun. Scientists who study the causes of climate change have long been interested in the variability of the Sun's radiation. We know that the visible part of the spectrum contains most of the energy we receive and that its variations are almost undetectable. Hence the term "solar constant" was invented to describe the Sun's output. The ultraviolet and x-ray parts of the spectrum contain only a small fraction of this output, but because it affects the photochemistry of the Earth's upper atmosphere and varies tremendously with solar activity, it is especially interesting to climatologists.

Precise measurements of the total amount of radiation arriving at Earth from the Sun (the solar "irradiance") began in 1980, with the launch of the Solar Maximum Mission in Earth orbit. It carried an instrument, the Active Cavity Radiometer Irradiance Monitor (ACRIM), capable of measuring the amount of energy reaching the Earth to one part in a million. Robert Willson, the inventor of ACRIM, noticed immediately that the rotation of large sunspot groups across the solar disk causes the irradiance to dip by as much as 0.2 percent. The amount of the deficit correlated with the total area of the sunspots and their relative darkness. Even a single emerging sunspot could decrease the light we receive. When an old sunspot decayed on the visible disk, though, the irradiance recovered over several days.

The Sun generates a steady flux of energy in its core, which must escape the surface eventually as radiation. If sunspots are darker than the average photosphere, it makes sense to look for areas that are brighter, where the deficit of radiation might escape. When observers looked for a bright ring around sunspots, however, they found none.

Hendrik Spruit, a Dutch astrophysicist, advanced a theory in 1977 and in 1982 to explain the absence of a bright ring. A strong field under a sunspot would deflect convection cells in its vicinity, as Biermann had suggested. Spruit estimated that the energy the cells carry could be stored as slightly hotter plasma throughout the full convection zone and would leak out as radiation only over many solar cycles. Peter Foukal, a Cambridge, Massachusetts, solar physicist, extended Spruit's theory of a sunspot in 1982 and predicted that the solar constant would vary over the solar cycle.

The plot thickened as the ACRIM data accumulated over a full solar cycle of eleven years. Willson discovered that the irradiance does indeed vary by about 0.3 percent over the cycle and is brightest when the number of sunspots reaches a maximum. That result was a surprise. It clearly implied that sources other than the photosphere must contribute to the irradiance. Two or more sources could seesaw in importance through the cycle, causing a net surplus or deficit.

What could these other sources of emission be? Foukal and his colleagues drew attention to the bright "faculae" that comprise the chromosphere of an active region. Could they partially offset the radiation lost in sunspots? Eugene Parker suggested that faculae are heated by some form of magnetic wave that is generated in sunspots, but a search for such waves was inconclusive. Perhaps such waves carry the energy even higher to the corona, where it is dissipated.

During the 1990s a small army of solar physicists bore down on the origins of the variability of the solar irradiance, not only for its intrinsic interest, but for the role it might play in climate change and in particular, global warming. In addition to sunspots and active regions, the strong magnetic fields at the borders of supergranules and between granules were shown to contribute somewhat to the modulation of the irradiance.

It now appears that surface heating in all these regions competes with the darkness of umbras in the variation of the irradiance. At sunspot maximum the faculae dominate; at sunspot minimum, the umbras do. We have advanced at least this far in our understanding.

But the Sun's magnetism affects the Earth in more ways than simply through irradiance. From 1645 to 1715, only about 50 sunspots were observed on the Sun, compared to a normal 50,000. During this period, called the Little Ice Age, Europe experienced an unusually cold climate. The Thames River froze over; snow fell in Rome; there was no summer warming to speak of.

Much has been made of this strong association between climate and sunspots since E. Walter Maunder pointed it out in 1893. It has stimulated research into the long-term fluctuations of the solar dynamo, which is presumed to be the ultimate cause of the dearth of sunspots. Theories also abound as to the possible physical connections between the Sun's magnetism and the Earth's climate.

One plausible scheme involves cosmic rays. As we shall see, the magnetic field carried in the solar wind shields the Earth from these energetic charged particles. As this field varies in strength during the sunspot cycle, the flux of particles reaching the Earth's atmosphere varies. Some researchers have suggested that cosmic rays influence the formation of clouds, a process that could seriously affect the climate of different regions of Earth, but the subject remains controversial.

So, what is the most important lesson we have learned? Our Sun is a variable star, not as dramatic as some, but a lot closer and therefore more relevant to our lives.

THE VIOLENT SUN

ON THE NIGHT OF DECEMBER 12, 2006, two astronauts, African American Robert Curbeam and a Swede, Christer Fuglesang, were installing a 2-ton truss between solar panels on the International Space Station. Suddenly they were told to scramble back on board the Space Shuttle *Discovery* just as fast as possible. A large solar flare had erupted just minutes before, and a blast of high-energy protons was expected instantly. They raced for cover but were caught outside before they could both enter. Each of them received a small dose of radiation, no worse than the amount they'd accumulate in a day's work in space. Curbeam would go on to log more than thirty-seven days of space-flight.

These astronauts were fortunate. This flare was rated as X3 because of its powerful pulse of x-rays, but happily it didn't emit a burst of 100-megavolt protons, which could have killed the astronauts. The flare was associated with the ejection of a massive cloud of coronal plasma, which caused a major geo-magnetic storm at Earth.

Solar flares are among the most dramatic events on the Sun. The amount of energy a big one releases is staggering. If we could capture it all (which of course is impractical) it could power the United States, at its present rate of consumption, for 100,000 years.

After a long debate, astronomers now agree on two basic types of flares, impulsive and gradual. In an impulsive flare an area on the Sun as large as ten Earths suddenly lights up. The flare seems to be confined to the low co-rona, often in a nest of loops that connect sunspots (fig. 4.1). Impulsive flares

Fig. 4.1. This active region was observed on June 6, 2000, with the Transition Region and Coronal Explorer (TRACE), as a flare was beginning. The dark filaments are magnetic loops in the low corona that connect opposite polarities. The loops are just beginning to flare. The sunspot on the left is about 20,000 km in diameter.

emit a pulse of radiation, lasting a few minutes and covering the whole electromagnetic spectrum, from gamma rays and x-rays to centimeter radio waves. When the x-rays strike our ionosphere, they can cause planetwide radio blackouts.

Plasma temperatures in such a flare can reach 30 million degrees, and streams of protons and electrons can be accelerated to relativistic energy. If these high-energy particles escape the flare site, they can damage satellite electronics and threaten the lives of astronauts.

A gradual flare, however, is the secondary effect of a larger catastrophe, a coronal mass ejection (CME). In such an event a large part of the corona is disrupted. First, a long horizontal ribbon becomes unstable and slowly rises. Then as much as a billion tons of magnetized plasma erupts into space at hundreds of kilometers per second, propelled by magnetic forces. An arcade of hot rising magnetic loops is left behind as the actual site of the flare, which may persist for hours as the "gradual flare." The loops drain into two bright parallel ribbons in the chromosphere. Such flares were therefore called two-ribbon flares. They are among the largest and most spectacular. Most coronal mass ejections are not associated with flares, though, and may even erupt far from any sunspots.

CMEs also produce interplanetary shocks, which can accelerate ions to megavolt energy. When a CME collides with the Earth's geomagnetic field, it can shut down electric power grids and cause widespread power blackouts. (We'll learn more about these effects in chapters 5 and 6.)

Solar physicists want to understand and be able to predict these violent events, not only because they affect us, but for their intrinsic astrophysical interest. Where does the energy come from? How is it stored for as long as a week and then released within minutes? How are protons and electrons accelerated to relativistic energy?

These are the basic questions that have intrigued solar physicists for the past forty years or more. Partial answers to all of these questions have been found from ever-improving observations, theory, and numerical simulations. In this chapter we'll review some of the highlights of what is known and what is still uncertain. We'll talk about impulsive flares and the way in which their energy is stored and released. Then in the second half of the chapter we'll move on to CMEs and the gradual flares they leave behind. We'll begin with active regions, where flares are spawned.

ACTIVE REGIONS

On June 6, 2000, the TRACE satellite captured images of two powerful x-ray flares that occurred in a small active region near the center of the disk. In figure 4.1, we see a snapshot of the region just as the flare is beginning. The active region consists of a nest of dark twisted loops that connect small sunspots. There is no doubt that we are looking at magnetic loops that have become unstable. Off on the left lies an isolated sunspot whose diameter, about 20,000 km, gives us some idea of the scale of the region.

Solar physicists have observed how regions like this develop on a blank solar disk. At first, a dark fibrous loop breaks through the photosphere and lengthens horizontally. Within an hour or so, tiny dark pores and small sunspots form at the ends of the filament. Within a day more loops and sunspots appear and the region continues to expand over the disk.

We now interpret this sequence as the emergence of a large omega loop rising through the photosphere, as we described in chapter 3. The sunspots and pores are located where the loop intersects the surface, but strong fields aren't limited to the sunspots. In the surrounding area and above, in the corona, fields as strong as 300 gauss may be found.

An active region like this one may join with others to form a "center of activity," which can persist for months. Fresh magnetic flux could continue to emerge sporadically. During such a period of growth, the center could spawn small flares (yielding 10^{19} to 10^{23} joules) several times a day, and one huge flare (say, 10^{25}, one followed by 25 zeros) perhaps once a month.

After a week or so, a typical active region begins to decay. Sunspots fade, perhaps by submerging; the region's magnetic fields spread out and weaken. The region is gradually dragged out in longitude by differential rotation. A diffuse bipolar region is left, with a marked neutral line that separates opposite polarities. A dark filament will often form on the neutral line, and as we shall see, can participate in a huge coronal mass ejection later on.

In figure 4.1, we can see that the flaring loops are twisting and writhing. These distortions are indications that the magnetic field is not "potential"; it must contain electric currents. Therein lies a tale.

The Preflare State

Long ago it was realized that the magnetic fields in an active region only store energy; they do not create it. The original source of energy for a flare is thought to lie in the deep convection zone, where huge convection cells rise and sink. These cells are presumed to interact with the roots of the fields we see above the surface. Recall that, according to Alfvén, the field is "frozen" in highly conducting plasma and must move with it. As the cells move about, they bend or twist the roots of an omega loop. High in the corona the magnetic loops react by twisting. These distortions increase tensions and pressures in the loops, which correspond to stored energy. This, in short, is the currently accepted view of the storage of flare energy.

Another way to look at the situation is to realize that twisting or bending a bundle of field lines generates an electrical current (see chapter 2 on Faraday's discovery of induction). From this viewpoint, it's the current that contains the stored energy. If the current decays or is interrupted, the field loses its twist by unwinding and returns to its minimum energy state, a current-free potential field.

Before it was possible to measure the magnetic fields in active regions, solar observers relied on images obtained in the light of the hydrogen alpha spectral line at 656.3 nm. They could see twisted and sheared fibrils in the chromosphere that strongly suggested nonpotential magnetic tubes. They could follow the relative motions of sunspots and pores that could be causing the distortions in the field. Even after Babcock invented the magnetometer, observers relied heavily on H alpha movies to follow the evolution of active regions.

Indeed, Harold Zirin, former director of the Big Bear Solar Observatory in California, famously said that H alpha photographs are the "poor man's magnetograph" because they are easier to obtain and show more detail than magnetic maps. Nevertheless, to confirm the scenario of flare energy buildup, everyone agreed that direct measurements of the field were needed.

By the late 1960s several observatories, including Mount Wilson, Kitt Peak, and the Crimean Astrophysical Observatory were making regular observations of magnetic fields and studying the development of active regions. These instruments had three serious limitations. First, they could detect only one component, along the line of sight, of the vector magnetic field. Second, measurements were limited to the photosphere, where the fields are strongest. And third, the spatial resolution of magnetic maps was poor compared to the sharp images observers like Zirin were getting.

Nevertheless, much was learned about the life cycle of active regions and the preflare state. Mere size and total magnetic flux were not good indicators of a region's propensity to flare nor the number and size of the sunspots within it. What seemed to matter was complexity, the interlacing of opposite polarities within a region. An extreme example was the "delta" type of sunspot, which had mixed polarities within its penumbra. It could produce especially energetic flares.

A region's rate of growth was another good indicator. When new magnetic flux continued to emerge rapidly, a region could produce barrages of flares over several days. Then after a pause, a region could recover and begin to flare again. It seemed as though the region had been refreshed somehow.

Observers found several types of "precursors" that indicated a flare might be imminent. In some cases, the long, dark H alpha filament in a region would begin to writhe or brighten. Or some preliminary bursts of x-rays and micro-waves might occur. Enhanced plasma flows or oscillations were also reported. None of these signs was an infallible predictor, however.

In the 1960s and 1970s, observers also looked for signs that the magnetic field of an active region either simplified or diminished as a result of a flare. They expected that a flare should drain some of the stored energy, and yet they found no evidence of that. Perhaps, they thought, the expected changes occurred higher in the corona, which was beyond their reach. Or more likely, the answers they sought lay in the components of the magnetic field they couldn't measure.

By the late 1970s a few groups turned their attention to building magne-tometers capable of measuring all three components of the field simultane-ously. The horizontal, or "transverse," components are intrinsically weaker than the vertical components and more difficult to detect. But after a decade of development, vector magnetometers were finally working reliably, although only at the level of the photosphere.

Mona Hagyard and her team at NASA's Marshall Space Flight Center were among the first to obtain interesting results with their instrument. (Hagyard is a small, pleasant woman whose soft southern voice belies her intense drive to compete. She retired recently after experiencing three complete solar cycles.)

Hagyard examined the so-called neutral line that divides regions of op-posite magnetic polarity at the photosphere of an active region (note 4.1). If the field were potential (that is, current-free), she would expect to see loops crossing the line at angles near 90 degrees. Instead, Hagyard and colleagues discovered that the horizontal fields near the neutral line were antiparal-lel on opposite sides of the line. In a word, the field was strongly sheared at the neutral line. That implied that a current sheet lies between the opposing fields.

Hagyard and friends went on to measure the amount of shear along the neutral line. They defined the shear as the angle between the actual direc-tion of a field line and the direction it would have in a potential field with the same vertical component. They determined that the shear increases before a flare, at least in some cases.

So they attempted to find the Holy Grail of flare observers, a reliable cri-terion for the onset of a flare, based on the amount of shear. They did find suggestive criteria, such as the minimum angle of shear and the length of the sheared zone along the neutral line, but these indicators often failed. Although shear seems to be necessary, it is not a sufficient condition for a flare. Some other trigger, perhaps the emergence of new flux with opposite polarity, also seems necessary to set off a flare.

Force-Free Fields

There is a curious aspect to the storage of energy. The magnetic field must be able to adjust to small, random distortions without disrupting. If each increment of energy were dissipated immediately, the region would never be able to collect the huge amounts it releases in flares. But the release requires some kind of instability. In short, the field has to be both stable to small injections of energy and unstable past a certain limit.

In 1954, Reimar Lüst and Arnold Schlüter, two scientists at the Max Planck Institute of Physics in Munich, wondered how this was possible. In thinking about the problem, they made a fundamental discovery about the stability of magnetic fields.

A potential field would remain stable indefinitely, but as soon as you twist it, a current is induced. Unless the current flows along the field lines, it will exert a so-called Lorentz force, which would disrupt the field rapidly. Since we don't see solar fields erupting all the time and since the plasma pressure in the solar atmosphere is far too small to balance the predicted Lorentz forces, Lüst and Schlüter concluded that the current always flows along the field lines. Such a field is called force-free (note 4.2).

We now think that all slowly changing fields on the Sun are nearly force-free. Obviously, flares would be exceptions. Lüst and Schlüter went on to show that in a force-free field the current is proportional to the local field strength. Moreover, the ratio of the current and the field, a quantity called "alpha," is constant along a field line.

Beginning around 1980, many theorists applied the increasing power of computers to investigate the storage of flare energy with numerical simulations. Among them were Zoran Mikic, Gerry Van Hoven, Terry Forbes, and their co-workers in the United States, and Eric Priest and his students at the University of Saint Andrews in Scotland.

In a typical two-dimensional calculation, they would begin with a potential field, an arcade or tunnel of magnetic loops that are anchored in the photosphere. Such arcades were thought to straddle the horizontal H alpha filaments that were often seen on the neutral line of an active region. Then they would subject the arcade to gross shearing motions parallel to the neutral line in the photosphere and follow the changes of the field geometry. At some critical value of shear, the field would become unstable, with a spontaneous formation of current sheets. The field might then erupt. Often the calculation would literally blow up at this point, with too rapid an evolution to follow. But these simulations helped to guide observers in searching for the relevant shearing motions.

In a parallel development theorists learned how to extrapolate actual observations of the photospheric vector field into the corona of an active region. Hermann Schmidt, a colleague of Schlüter, had shown how to compute a three-dimensional potential field as early as 1964. In the mid-1980s, Mikic and others developed methods to compute force-free fields. When they compared their computed loops with the observed H alpha structures, they found good agreement, at least in selected cases. This was an important step toward understanding how active region fields are shaped, but it didn't lead to a better criterion for the onset of a flare.

Both observations and numerical methods have improved significantly since these early efforts. Models of coronal fields that are extrapolations of photospheric observations remain one of the principal tools for studying how active regions store energy.

The Buildup of Flare Energy

Since the 1980s, many observers have tried to follow the buildup of magnetic energy in an active region and to determine when and why it finally flares. There is a vast literature on the subject. Shear has remained an interesting indicator, but other factors look more promising.

In 1989 Richard Canfield and his colleagues at the University of Hawaii introduced the idea of monitoring the electric currents that flow in and out of an active region at the photosphere. They reasoned that, if a field stores energy in electric currents, they should be able to detect a significant change before or after a flare.

Canfield and company used the magnetometer at the Mees Solar Observatory in Maui, Hawaii, to measure the vector fields in an active region over a period of five days. From the horizontal components of the field, they were able to construct maps of the vertical currents at the level of the photosphere.

Let's take a moment to understand how they did it. Imagine a twisted bundle of field lines poking straight up through the photosphere. The horizontal components of the field will look like little arrows that march around a circle centered on the axis of the bundle. By Ampère's rule (see chapter 1), the strength of this horizontal component is proportional to the strength of the current flowing along the vertical component of the field. So, by measuring the horizontal components of the vector field at a grid of points within the area of a region, one can estimate the distribution of currents, at least in principle.

The region Canfield and team picked produced a few small impulsive flares but no marked variations in the currents before or after the flares. This result might have been a disappointment except that it disproved one of the oldest proposals for the onset of a flare, the so-called current interruption model, attributed to Alfvén and Carlqvist (note 4.3).

In 1995, Canfield and his colleague Andrei Pevtsov introduced another measurable quantity that promised to indicate the growth of nonpotential energy in an active region. This variable is the helicity, the ratio of the vertical current to the vertical component of the magnetic field. The helicity is a measure of the amount of twist in a magnetic tube. As we shall see, it became the focus of a considerable debate later on.

Pevtsov computed the distribution of helicity in each of sixty-nine active regions, a heroic job. He found that each region was speckled with a fine-grained pattern of helicity of both signs. When he averaged the helicity measurements over a whole region, he found a small net helicity that persisted for at least a day. In the northern hemisphere the net helicity was negative for a majority of cases; in the southern hemisphere the net helicity was positive. The hemispheric variation suggested a weak influence of the Coriolis effect and of differential rotation. But once again, changes of the net helicity did not prove to be an indicator of impending flares.

Several groups have followed up on the pioneering work of Canfield and colleagues, to try to detect changes in electric current or helicity, and hopefully, find a critical threshold. Yuanyong Deng and the staff at the Huairou Solar Observing Station in China had the good fortune to measure vector fields in the active region that produced the famous Bastille Day flare in

2000. This two-ribbon flare had it all: 50-kiloelectron-volt (kev) x-rays, 4-megaelectron-volt (MeV) gamma rays, 30-MeV protons, 30 million–degree plasmas and a huge coronal mass ejection.

Unlike Canfield and company they did see systematic changes of the current and helicity distributions in the days before the flare but no "smoking gun" to indicate that some threshold had been crossed just before the flare. Other signs of the growth of magnetic energy were observed, including changes in the shear of horizontal fields. The shear was especially noticeable along the neutral line. A long dark filament, visible in the H alpha line, lay along the neutral line and played a key role in the coronal mass eruption that occurred later.

Yihua Yan and colleagues at the Huairou Station also observed the events leading up to the Bastille Day flare. They used their vector field measurements to calculate the 3-D coronal structure of the region. They were able to reconstruct the twisted rope that coincided with the H alpha filament on the neutral line. Later this rope erupted as part of the CME. Above the rope their calculations yielded a nested set of magnetic arcades, each of which was sheared at a different angle with respect to the neutral line. Their results strongly resemble the arcade observed by TRACE just after the flare occurred.

In the best of all possible worlds, such 3-D models might enable one to detect the signs of an impending disequilibrium of forces or a significant change in the current distribution. So far, alas, this has not been possible. We understand how active regions store energy and how their fields are arranged but not why and when they flare.

THE PROBLEM OF ENERGY RELEASE: RECONNECTION

At some point the magnetic field in an active region becomes unstable and dumps its energy as a flare. Energy is often released in two distinct phases. In the impulsive phase, which lasts a few tens of seconds, the flare emits a small fraction of the stored energy as high-energy electrons and protons. These energetic particles then produce hard x-rays, gamma rays, and microwave radiation. In the gradual phase that follows, which may last for hours, the bulk of energy is radiated as soft x-rays.

Solar physicists understand the gradual phase reasonably well, because it involves a familiar process, the cooling of extremely hot plasma. The impulsive phase, on the other hand, has challenged them for decades, because it involves a variety of nonthermal processes. Two questions have been particularly difficult to answer: How can the magnetic field release its energy so quickly? And how can charged particles accelerate to relativistic energy in so short a time?

Their best answer to both questions so far is "reconnection," the severing and restructuring of the coronal field in an active region. It is possible only if the plasma possesses at least a minimum resistance to electricity. Reconnection simplifies the original field and releases some of its free energy. In an extreme case, antiparallel fields may actually annihilate each other. Reconnection is thought to occur in many astrophysical situations, as we'll see in later chapters, so we'll describe it in some detail here.

Ronald Giovanelli, the Australian physicist whom we met in chapter 3, introduced one of the essential elements of reconnection in 1947 (note 4.4). He realized that strong electric currents could be induced by the changing magnetic fields around a sunspot. If a neutral point existed in the field, where the magnetic field vanished, a complete circuit could be closed. A sustained electric field would then accelerate electrons to high energy, and their collisions with atoms would produce the emissions he saw in flares.

Paul Sweet, a physicist at the University of London Observatory, picked up the idea in 1958. He investigated the collision of two pairs of sunspots that are driven together by subsurface forces. Antiparallel field lines in the space between the pairs would meet in a neutral sheet, only a few meters thick, where they could reconnect into hairpin shapes and release a part of their energy (fig. 4.2A). In the process a strong electric current would be induced in the sheet. The colliding fields would continue to push plasma into the neutral sheet, where the electric current would heat it. Hot plasma jets would squirt out of the ends of the sheet at high speed, carrying the reconnected field lines. Sweet identified the jets as the flare we observe.

Eugene Parker, the physicist we have met in chapters 2 and 3, spotted a flaw in Sweet's argument. He pointed out that antiparallel field lines are always embedded in plasma. For the lines to reach each other and cancel, the plasma they carry must diffuse into other lines through a thin interface. Diffusion of charged particles is very slow across field lines, however, and therefore it limits the rate of energy conversion (note 4.5). Pushing the fields together helps to speed up the conversion, as Sweet suggested, but you can't push the fields together faster than diffusion permits.

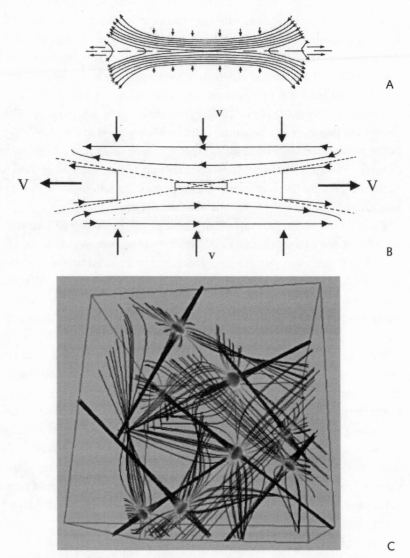

Fig. 4.2. Three models of magnetic field reconnection are shown here. A: the two-dimensional Sweet-Parker model; B: the Petschek model; C: the three-dimensional Priest-Forbes model.

Parker estimated that antiparallel fields of, say, 500 gauss and 10,000 km long, would require hours or days to release the amounts of energy observed in flares. That result conflicted with the tens of seconds of the impulsive phase. Parker concluded in 1962 that the annihilation of magnetic fields failed the test of observation and that an alternative cause should be sought.

Two years later, a fresh idea surfaced that rescued the reconnection scenario. Harry Petschek, a geophysicist at the Avco Everett Research Laboratory, proposed that standing shock waves would develop at the ends of a very short neutral sheet (fig. 4.2B). In his scheme, only short lengths of field lines need to diffuse through the sheet in order to reconnect and liberate magnetic energy. Most of the plasma avoids the bottleneck in the diffusion zone and heats up by passing through the shocks. The result is a much faster inflow speed and a very much faster rate of energy conversion than is possible in the Sweet-Parker scenario. The maximum rate was still not fast enough, but at least a path for further improvement was opened.

For a time, Petschek's scheme seemed to solve the problem of fast energy release. Then problems arose. Nobody could reproduce his scenario in numerical simulations. The flux would pile up at the boundary of the interface if it were pushed in faster than diffusion allowed. Laboratory experiments also showed that energy conversion occurred at rates no faster than predicted by Sweet and Parker.

Plasma physicists stepped in to offer suggestions. The key to the situation was in the neutral sheet, where reconnection takes place. Perhaps the classical theory of diffusion was at fault. Diffusion depends on the same collisional processes that create the electrical resistance of plasma. So if one could find mechanisms to enhance resistivity, they might also accelerate diffusion.

Plasma physicists had known about "anomalous resistivity" for decades. They were engaged during the 1970s and 1980s in trying to produce a controlled thermonuclear fusion reaction in a laboratory. In principle such a conversion of nuclear energy to useful heat could solve the world's energy needs, with none of the radioactive by-products produced in conventional fission reactors.

These researchers tried to contain the extremely hot plasmas in their laboratory machines (tokamaks) with a carefully designed magnetic "bottle." When they heated the plasma with electric currents, they discovered that the plasma's electrical resistance could suddenly increase by many factors of ten. As a result, magnetic field lines could diffuse rapidly through the plasma and reconnect. The bottle was then no longer closed, and the plasma could escape to the walls of the machine, with catastrophic effects (note 4.6).

Anomalous resistivity involves a great variety of subtle plasma processes and is still not completely understood. In general it arises when electric currents exceed the carrying capacity of plasma. At that point, the plasma becomes highly turbulent. Moreover, a host of exotic magnetic and electrical waves are generated that interfere with the conduction of current. Both of these factors increase the plasma resistivity and therefore the diffusion of fields through plasma in a reconnection zone.

Solar physicists invoked anomalous resistivity to improve the Petschek model of reconnection, but the predicted rates of energy conversion were still too slow to explain the fastest events in the impulsive flare.

Harold Furth, John Killeen, and Marshall Rosenbluth outlined a theory of the reconnection of magnetic fields that changed the nature of the problem. They calculated that pressing antiparallel fields together would result in a spontaneous instability. The parallel fields would dissolve into a chain of circular loops in a process they called the tearing mode. It proceeds at a much faster rate than any previously proposed and was immediately adopted by solar physicists.

In the forty years since the introduction of the tearing mode, reconnection has been studied intensively in the laboratory and in numerical experiments. A zoo of complicated plasma processes seems to flourish in the enigmatic neutral sheet, and at this time there is no established theory applicable to solar flares. As a result, many solar physicists have fallen back on some form of the Petschek theory to interpret their observations. Saku Tsuneta, from the University of Tokyo, for example, posited reconnection at the top of a set of coronal loops, complete with slow shocks, to explain the complicated events in his famous flare of 1995.

Until recently, only the two-dimensional cases illustrated in figure 4.2A and B were investigated. In the past six years, Eric Priest and Terry Forbes have tackled the more realistic situation in three dimensions. As you might expect, they've discovered even more complexity.

Priest is an inspiring teacher who has trained a generation of mathematical physicists at the University of Saint Andrews. His specialty is hydromagnetics, the interaction of magnetic fields and plasmas. With his longtime collaborator Forbes, at the University of New Hampshire, he's been investigating the intricacies of solar activity for three decades. Priest bounces with energy and enthusiasm. He and Forbes published a monograph on reconnection and its astrophysical applications in 2000 and are writing another about the three-dimensional case.

Priest has cautioned, "Studies of reconnection in three dimensions are still in their infancy." But already he and his colleagues have discovered that a neutral point in a 3-D field has two key elements: a spine (a field line) and a fan of field lines intersecting the spine (see fig. 4.2C). Reconnection may occur between spines, or between fans, or at the intersections of the fans of two neutral points, or in the layer that separates two fields of different topology. The possibilities are fascinating but make the mind spin. We'll have to leave the subject here.

ACCELERATION OF CHARGED PARTICLES

Next we come to the most difficult problem theorists face in explaining the behavior of impulsive flares: How are charged particles accelerated to high energy in so short a time and in so limited a space?

Observations from a series of satellites, beginning with the Solar Maximum Mission (1980) and continuing with the Ramaty High Energy Solar Spectroscopic Imager (2002), have defined the scope of the problem. They reveal that nearly the entire population of electrons in a flaring loop (10^{37} particles) accelerates to energy as high as 100,000 electron volts within one or two minutes. Some must reach this level within a fraction of a second to account for hard x-ray spikes.

The bulk of fast electrons spiral down the field lines in the legs of the loop and emit polarized (gyrosynchrotron) microwaves along the way. When the electrons smash into the chromosphere, they create hard x-rays by impact with ions. They also heat the plasma there instantly to tens of millions of degrees so that it explodes up the legs of the loop.

For a while, scientists debated whether some of these fireworks were not emitted by superheated plasma, say, at a temperature of 10^8 degrees. Recent observations favor the alternative that nearly all the flare energy is released as the kinetic energy of fast electrons.

Ions are also accelerated in an impulsive flare to energy as high as 100 MeV. Some of these escape from the flaring loop and are detected in interplanetary space. The rest, mainly fast protons and helium nuclei, collide with heavier ions, such as carbon, oxygen, and neon and excite their nuclear energy levels. These heavy ions then radiate a blizzard of gamma ray photons, with energies between 1 and 10 MeV. Positrons and electrons also annihilate to produce the telltale gamma ray at 0.511 MeV.

Theorists like James Miller and Gordon Emslie at the University of Alabama, Peter Cargill at Imperial College, London, and their co-workers have thought hard and long about the processes that accelerate flare particles. They recently summarized the best candidates, but many details remain unsettled.

Flare particles can be accelerated by at least three different mechanisms: sustained electric fields, shocks, and hydromagnetic waves. It should be no surprise that all three are likely to operate during reconnection of a magnetic field. When magnetic fields collide at a thin plasma layer, a strong electric field is induced, which can drive a heavy electrical current. An electron in the stream can pick up sufficient speed in a few milliseconds to enable it to avoid collisions with slower particles in its path. Then there is nothing to prevent it from retaining all the energy the electric field injects. It literally "runs away." If the reconnection lasts for a few seconds, the electron can pick up tens of kilovolts of energy.

In some models, standing shock waves enclose a reconnection site. A shock wave is a thin, turbulent layer that separates different plasma flows. The plasma speed, density, and temperature change by large amounts across the shock. In passing through a shock, both ions and electrons can collide repeatedly with moving blobs of magnetized plasma, or "scattering centers." They pick up energy at each collision. Shocks propagating away from the flare can also push particles to high energy.

Finally, reconnection is thought to generate a great variety of waves that can interact with charged particles. A particle can gain energy in at least two ways: by random (stochastic) contacts or by resonant coupling. In a turbulent region, where waves of many frequencies are propagating, a particle may gain or lose energy at each encounter and still experience a net gain. This is stochastic acceleration. In a resonant encounter, the particle's rate of spin around a field line (gyrofrequency) matches the wave frequency and the particle gains energy.

Perhaps the weirdest aspect of the impulsive flare is the preferential acceleration of heavy ions, particularly helium. Helium has two isotopes: helium three, which has two protons and one neutron; and helium four, which has two neutrons and two protons. In the quiet corona, helium four outnumbers helium three by a factor of 2,000. But the two isotopes are equal in number among the megavolt ions emitted in the impulsive phase.

Two plausible explanations have been proposed. The gyrofrequency of an ion depends upon its mass; heavier ions spin more slowly in a magnetic field. So if a heavy ion intercepts an ion-cyclotron wave of the same frequency, it can gain energy selectively. Alternatively, an electron beam can generate Alfvén waves under certain circumstances, and these can accelerate ions of the same gyrofrequency. The beam model can account for preferential acceleration, not only of helium three, but also of neon, magnesium, silicon, and iron ions.

CMES AND TWO-RIBBON FLARES

We turn now to one of the most spectacular events on the Sun, the coronal mass ejection and its associated two-ribbon flare. The Bastille Day X5.7 flare of July 14, 2000, is a fine example. The event occurred in an active region at the center of the solar disk and began with the eruption of a long horizontal filament in the low corona. The overlying corona lifted off with it, at a speed of about 1,600 km/s, aimed directly at Earth. An interplanetary shock wave, signaled by a decametric Type II radio burst, preceded the CME. It accelerated protons in interplanetary space to 10 MeV. The hail of protons reached the Earth two hours after the start of the flare and created gamma rays in the atmosphere. Some twenty-two hours later, the CME arrived at Earth and produced the largest geomagnetic storm since 1989.

Figure 4.3 shows a similar CME as observed in white light by the Large Angle and Spectometric Coronagraph (LASCO) aboard the SOHO satellite. The small white circle at the center represents the size of the solar disk, so you can gauge the enormous size of such an eruption. (The dark disk at the center is the part of the coronagraph that eliminates the bright solar disk from the image.) Observers estimate the CME mass at a billion tons of plasma.

As the Bastille Day CME erupted, a pulse of hard x-rays and gamma rays flashed out, illuminating a pair of bright parallel ribbons in the low corona.

Fig. 4.3. This "Light Bulb" coronal mass ejection was observed by the LASCO coronagraph aboard the SOHO satellite on February 27, 2000. The small circle at the center outlines the solar disk. Three typical parts of a CME are shown: a bright ring that outlines an expanding shock; a "void"; and an erupting filament.

Analysis of this radiation, and images in hard x-rays, indicated that electrons with energy up to 90 keV and protons with energy to 10 MeV had bombarded the chromosphere. A scorching stream of fast particles lasted for several minutes.

Solar physicists have observed such dramatic two-ribbon flares since the 1960s. Peter Sturrock, a professor at Stanford University, proposed the first viable model for such an event in 1968. Sturrock's specialty is plasma physics, and he has been at the forefront of solar physics for almost forty years. As we shall see, he has many other interests in plasmas and fields throughout the universe.

Sturrock got the idea for his model of a flare when he saw a white-light photograph taken at the total solar eclipse in 1965, which showed a cross-section of a helmet streamer. Streamers are gigantic sheets of luminous plasma that extend tens of solar radii into space. They stand astride the neutral lines of old decayed active regions, where long horizontal filaments form. You can see a few of them in figure 4.3 (diametrically opposite the CME) and in figure 5.1.

Sturrock's concept of the coronal magnetic field in a streamer is shown in figure 4.4 (top). In this sketch we are looking at the cross-section of a long, fan-like structure that continues into the page. Closed field lines arch across the neutral line of a bipolar magnetic field. Higher in the corona the field lines open to outer space. Sturrock guessed that the recently discovered solar wind had drawn out the field lines.

He postulated that antiparallel fields in the high corona meet in a "sheet pinch," a metastable thin layer that stores the energy injected by the solar wind. The sheet eventually disrupts spontaneously by the tearing mode, beginning at the neutral point, where the sheet meets the topmost closed field line. Field lines reconnect as shown in figure 4.4 (bottom), and propel a massive cloud outward. Charged particles accelerate in the reconnection and shower down the legs of the structure to illuminate two bright ribbons in the chromosphere.

Sturrock's model gave a qualitative explanation for many of the features observed in the large eruptive flares that accompany CMEs. It has been developed further by many authors and applied by such solar physicists as Saku Tsuneta to interpret the complex events they saw in Yohkoh satellite and SOHO observations. But this model was almost too successful; it became the paradigm for the origin of CMEs for over twenty years. Solar physicists were persuaded that CMEs are preceded by two-ribbon flares.

In parallel with this development, another community of scientists was studying the interplanetary effects associated with two-ribbon flares. Only after the Skylab mission of 1973–1974 established that CMEs are the real

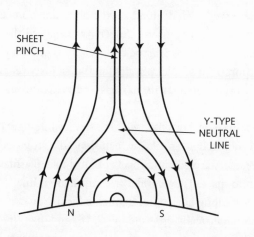

SHEET
PINCH

Y-TYPE
NEUTRAL
LINE

N   S

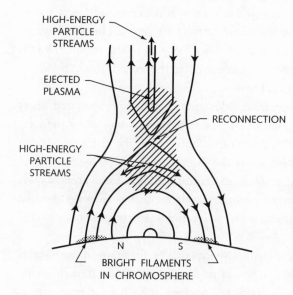

HIGH-ENERGY
PARTICLE
STREAMS

EJECTED
PLASMA

RECONNECTION

HIGH-ENERGY
PARTICLE
STREAMS

N   S

BRIGHT FILAMENTS
IN CHROMOSPHERE

Fig. 4.4. Peter Sturrock's model of an eruptive flare is shown in two stages. At the top is a magnetic structure similar to a helmet streamer in cross-section. At the bottom the neutral sheet at the centerline reconnects, initiating a chromospheric flare and a coronal mass ejection.

sources of large geomagnetic storms, did the two communities begin to exchange ideas to any great extent. In 1993 a controversy broke out between them.

John Gosling triggered the debate. Gosling was working at the Los Alamos National Laboratory near Santa Fe, New Mexico. For many years he'd been studying high-speed streams in the solar wind; interplanetary shocks;

and CMEs. In 1976, he was among the first to identify CMEs as the cause of large geomagnetic storms. When he spoke out, people listened. He was blunt but usually on target.

In 1993, he published a paper called "The Solar Flare Myth" that irritated the tight community of solar physicists. It also stimulated a reassessment of the relationships between flares and CMEs.

Gosling pointed out that for many years, observers had seen large flares that preceded severe geomagnetic storms, shocks in the solar wind, and blasts of energetic particles in interplanetary space. Therefore, the notion that flares cause these effects took firm root in the minds of solar scientists and remained the conventional wisdom for twenty years. But when one studies a larger body of data, he wrote, one arrives at a completely different conclusion: the CME is the primary event, and the flare is a by-product.

Steve Kahler at Boston College had laid out the case in a scholarly article the previous year. He pointed out that CMEs erupt with speeds as slow as 50 km/s and as fast as 1,200 km/s. The slow CMEs don't produce interplanetary disturbances or large geomagnetic storms. They may or may not be accompanied by a flare.

In contrast, the fast CMEs are less frequent, but they cause the biggest geomagnetic storms. They are usually preceded by the eruption of a filament, not by a flare. Only after the eruption has begun does an impulsive flare appear as two parallel x-ray ribbons in the chromosphere. As the ribbons separate horizontally, a series of high "postflare" loops form in the corona, which continues to radiate x-rays for hours. These so-called two-ribbon flares are physically distinct from the low, contained flares that occur in strong magnetic fields.

Interplanetary observations confirm the distinction between flare-generated and CME-generated disturbances. Only fast CMEs accelerate heavy ions in the solar wind to megavolt energy. Moreover, the chemical composition of the plasma in CMEs differs from that in the ejections of flares. (We'll return to this subject in the next chapter.)

Gosling's article roiled the solar flare community and evoked some heated rebuttals. But even traditional solar physicists have come to accept his conclusions. They realized that workers with different specialties could view a complex event very differently. Nevertheless, one wonders how so many smart people held the wrong opinion for so long a time?

There are several good reasons. Flares, even two-ribbon flares, have been relatively easy to observe in H alpha for decades and in x-rays for at least twenty years. CMEs in contrast are difficult to observe. One needs a special telescope (a coronagraph) that masks the bright disk of the Sun and allows the fainter CME to be seen beyond the mask. Astronomers were able to study these events in detail only after coronagraphs flew aboard Skylab. Only with the launch of the Yohkoh satellite in 1991 was it possible to observe the onset of a CME in x-rays on the disk of the Sun.

The real source of confusion, though, was the great variety of associations that exist among flares, filaments, and CMEs. It required careful analysis of a huge database to arrive at what we hope is a clear picture of the CME phenomenon.

Solar physicists now recognize that impulsive flares and CMEs release energy differently. In an impulsive flare, reconnection releases the bulk of the energy primarily as the kinetic energy of fast-electron and ion beams. Their energy dissipates as heat and radiation, usually without any eruption. In a CME, most of the energy is released as the kinetic energy of a large mass. Reconnection aids the escape of the mass and produces the high-energy sideshow but does not propel the mass. Instead magnetic forces propel the mass.

Beginning in the early 1980s theorists began to investigate how a CME might develop from a stressed coronal field. Boon Chye Low, at the High Altitude Observatory, was among the first. Low was a student of Eugene Parker and has probably surpassed his mentor in his ability to pose and solve difficult three-dimensional problems in the dynamics of magnetic fields.

In 1981 Low demonstrated that a force-free field is able to expand slowly under the influence of motions at its footpoints in the photosphere. But a critical point is reached beyond which no stable equilibrium is possible. Low suggested that at that point the field would erupt catastrophically.

The following decade saw many improvements in our understanding of solar catastrophes. In 1990, Eric Priest and Terry Forbes took an important step forward. They showed how photospheric flows that converge on a magnetic arcade could generate a helical coil under the arcade. They identified the coil with an H alpha filament.

As the flows continue to compress the legs of the arcade, energy is stored in the coil, it rises, and a vertical neutral sheet forms beneath it. The coil rises and presses against the overlying arcade. The system slowly swells in size, as energy is stored. The authors assume, then, that the sheet eventually reconnects, allowing the filament to erupt. The arcade would be torn apart catastrophically.

They couldn't follow the eruption in detail, but this was a good first step. It pointed to the formation of a coil that resembles the filament whose eruption initiates a CME. Reconnection would only assist in freeing the CME, but would not provide the energy to drive it.

Then in 1991, French astrophysicist Jean-Jacques Aly and Peter Sturrock at Stanford University proved a theorem that upset the whole idea. They showed that any closed force-free field whose footpoints are anchored firmly in the photosphere contains less energy than the same field after it has opened to space. That would mean, for example, that an arcade or a filament could never escape the Sun. How depressing!

Several groups have since found different ways around this obstacle. We'll discuss two of them briefly. Tahar Amari and his colleagues at the Center for Theoretical Physics in Paris demonstrated a three-dimensional model in 2000, in which reconnection near the neutral line cuts the "tethers" that hold down a rising magnetic coil, rather like releasing a balloon.

They started their simulation by shearing a potential field arcade, using antiparallel motions in the photosphere along the neutral line. The shearing generated a twisted coil within the arcade (fig. 4.5 left). The coil strains to erupt, but the overlying arcade restrains it. At this point, Amari and friends propose, the feet of the arcade come together at the photosphere. The tethers that hold down the coil are cut, and the coil is free to explode into the corona (fig. 4.5 right). The reconnection also reduces the energy of the final open field below that of the coil, thereby avoiding the Aly-Sturrock constraint. Sufficient magnetic energy is released to propel a large mass at high speed.

In a parallel development, Spiro Antiochus and friends at the U.S. Naval Research Laboratory have developed a "breakout" model over the past seven years. Antiochus was a former student of Peter Sturrock. He now leads a group of theorists who have made important contributions to solar theory.

This group's scenario starts with two bipolar regions parallel to each other. A magnetic arcade lies between the regions. Shearing motions are then applied to the two sides of the arcade, which causes the arcade to push outward. The large-scale field, which overarches both bipolar regions, resists the arcade's rise. But when reconnection occurs above the arcade, these restraints

Fig. 4.5. A twisted flux rope erupting into the corona as a CME. The dark patches at the photosphere are regions of opposite magnetic polarity.

vanish and it begins to break free. Then reconnection occurs beneath the arcade, forming a twisted coil. As it rises, the large-scale field above it collapses, propelling the coil ever more rapidly.

These models explain many of the features actually observed in CMEs. Several other groups have proposed similar models that do equally well, but they all invoke large-scale persistent flows at the photosphere to drive the corona to the point of catastrophe. Observers have looked for such flows by tracking either patches of field or granules. Flows are certainly found, but they are, for the most part, random and episodic. Where else could one look?

The answer may lie deeper in the convection zone. There, large cells of plasma are predicted to rise and turn over slowly, over a period of weeks or months. They would have the power to distort the roots of the fields we see in the corona and drive them to instability. Is there any evidence this is actually happening?

We still can't observe strong magnetic fields or flows very far under the photosphere, but some progress has been made in recent years. Thomas Duvall at the NASA Laboratory for Astronomy and Solar Physics, Alexei Kosovichev at Stanford University, and their co-workers developed the necessary techniques during the late 1990s. They use SOHO observations of acoustic waves at the surface (seismic tomography) to derive the circulation of plasma beneath it.

Recall that we described in chapter 3 how such methods have revealed the temperature structure and flows underlying sunspots. Kosovichev has also applied the technique to map the flows as deep as 20,000 km below an active region. SOHO provides data continuously, and a region could be followed for at least a week as it crossed the disk of the Sun. We shall have to wait to see whether long-lived cells can be discovered.

We've come a long way toward understanding the flare phenomenon, but the key process of reconnection is still an open frontier. Similarly, the persistent flows that are postulated to store energy in a coronal field are still somewhat elusive. A combination of improved observations, numerical simulations, and insights from the plasma physics community should help solar physicists to crack this hard nut in the future.

In the next chapter we'll continue the story of coronal mass ejections by following them out into interplanetary space. Looking back from there to the corona has yielded some deep insights into the nature of solar activity.

THE HELIOSPHERE

Winds, Waves, and Fields

As we go about our daily routines on a sunny, spring-like day, we're blissfully unaware of the storm that rages just outside the Earth's atmosphere. A supersonic wind of plasma blows off the Sun in all directions, past the Earth and beyond Pluto. The wind is strong enough to sweep back the Earth's magnetic field, like the streaming hair of a girl on a motorcycle.

We've known that the wind exists only since 1962. Since then we've learned a lot about its behavior from a series of satellites and probes, culminating in the ongoing Ulysses mission. But scientists still don't agree on just how the wind reaches speeds as high as 800 km/s. That question remains one of the most challenging that theorists still face. Magnetic waves are probably responsible, at least in part.

We do know that the wind originates in the solar corona. But the corona is also something of a mystery, despite decades of improving observations from satellites. The corona is millions of degrees hotter than the surface of the Sun, an anomaly that stunned astrophysicists when they learned about it in the 1940s. Many ideas have been proposed to account for the high temperatures, but none has won out as yet. Once again, magnetic waves are suspected to be responsible, but there are other proposals on the table.

The high speed of the wind and the high temperature of the corona are two parts of a larger problem, the origin of the heliosphere. The word "heliosphere" is relatively new in astronomical jargon. It refers to the gigantic bubble that the solar wind sweeps out of the interstellar gases. The bubble extends to twice the distance to Pluto and possibly farther. Somewhere out there the wind collides with interstellar clouds in a so-called terminal shock.

In this chapter we'll recall how we discovered the high temperature of the corona, and we'll survey its magnetic structure. Then we'll discuss some proposals for heating it. Next we'll turn to the discovery of the wind and a quick look at fast and slow wind streams. Finally, we'll look at ideas to account for the high wind speeds.

A HOT CORONA

The temperature of the corona has a long and interesting history. In the nineteenth century the corona could be seen only during total eclipses of the Sun. Astronomers were amazed to see long "streamers" extending a solar radius or more beyond the edge of the Sun, like the petals of a daisy (fig. 5.1). At the time, the corona was thought to be cooler than the photosphere, because the temperature was known to decrease outward in the photosphere. A cool corona could not extend very far into space. So the existence of these long streamers posed something of a problem.

Walter Grotrian, astronomer at Potsdam Observatory, was the first to question a cold corona. He had observed an unidentified line in the spectrum of the corona during a solar eclipse. Princeton astronomer Charles Young had discovered the line at 530.3 nm back in 1869. Because it had never been seen in the laboratory, he had assigned it to a new element he called coronium.

In the meantime, Caltech astronomer Ira Bowen had shown that ordinary elements such as oxygen and nitrogen emit unfamiliar lines in very low-density gases. He correctly identified several problematic lines that had been attributed to another ad hoc element, "nebulium." Grotrian thought that Bowen's theory might apply equally well to the corona. Perhaps a familiar element was radiating Young's line because of some strange conditions in the corona. What could these be?

Meanwhile, Swedish physicist Bengt Edlén had produced a bonanza of previously unknown spectral lines in the laboratory by exposing gaseous elements to an extremely hot electrical spark. In 1939, Grotrian used Edlén's data to show that several unidentified lines in the corona could have been emitted by highly ionized iron. Young's "green" line at 530.3 nm, for example,

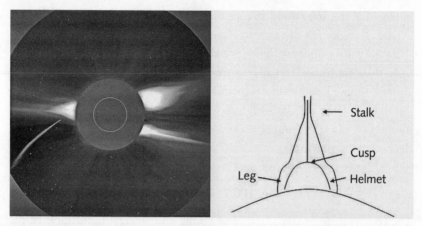

Fig. 5.1. Helmet streamers located near the solar equator around sunspot minimum are shown in this image from the Large Angle Solar Coronagraph aboard SOHO. Specific parts of a streamer are shown in the sketch.

might originate from iron atoms that have lost thirteen of their twenty-six electrons. For iron to exist in such a state, the coronal temperature must be as high as 2 *million* degrees. Other coronal lines yielded ionization temperatures of 1 million degrees. The corona was *hot*.

Soft x-ray observations from rockets and satellites yielded the first images of the corona over the solar disk. At wavelengths between 150 and 30 nm, there are spectral lines that shine only in narrow ranges of temperature between 100,000 and 3.5 million degrees. Images obtained in such lines revealed where the corresponding temperatures prevailed. Active regions, for instance, possess plasma as hot as 2.5 to 3.5 million degrees, while quiet regions top out at about 1.5 million.

Why Is It So Hot?

The search for a mechanism that could heat the corona to millions of degrees began soon after Edlén's work became known. Ludwig Biermann (a familiar name by now) and Martin Schwarzschild (an astrophysicist at Princeton University) proposed in 1948 that sound waves generated in the convection zone would travel to the corona, where they would dissipate and heat the plasma. Since then, their idea has been explored to exhaustion by a series of theorists.

Peter Ulmschneider at Göttingen University carried out extensive calculations throughout the 1980s. He learned that sound waves steepen into shocks in the high photosphere or chromosphere and dissipate there. That was encouraging, because that is where the dense plasma radiates strongly and where most of the energy supply is needed. But whether a small portion of the energy could reach the corona depended on many unknown factors. The frequency and energy flux of sound waves was quite uncertain, for example. Moreover, theories of the generation of sound waves in the convection zone were rudimentary at best, and observations in the 1980s only confused the issue with many false positive identifications of sound waves. Until recently, sound waves were not thought to be important in heating the corona. As we shall see, the idea has been revived.

As sound waves lost their luster, solar physicists turned to magnetic fields. They recognized that the strong magnetic fields in active regions could play some role in heating them, and they began to explore ways in which the fields could channel energy from the convection zone to the corona. The basic idea is that turbulent convective motions twist the subsurface roots of the coronal field lines and generate waves or electric currents that heat the corona.

The scientists focused on two basic types of processes that differ by the speed of the driving motions relative to the Alfvén speed (note 5.1). *Slow* motions would generate DC electric currents along the field lines, which could heat the coronal plasma like the coils in a toaster. Alternatively, slow motions would distort the coronal field and create current sheets, which could dissipate energy by magnetic reconnection. *Fast* motions, on the other hand, could generate a variety of magnetohydrodynamic (MHD) waves that might propagate to the corona and dump their energy there.

Steady electric currents were first proposed by Eugene Parker in 1972. He visualized slow photospheric motions "braiding" the field lines and generating currents. Later it was realized that the currents were unlikely to remain steady; they would degenerate rapidly by the tearing mode we discussed in chapter 4 in connection with flares. So until recently, the braiding of field lines was not considered a mechanism distinct from reconnection. In 2004, however, three Swedish theorists constructed a three-dimensional dynamical model of the process (Hardi Peter, Boris Gudiksen, Äke Nordlund, *Astrophysical Journal* 61, no. 7, L85, 2004). The coronal spectrum predicted by their model agrees tolerably well with observations. So this heating mechanism still seems to be feasible.

Heating by some form of reconnection has seemed very promising for a long while. Eugene Parker proposed in 1988 that, as the coronal field is slowly distorted, reconnection sites appear randomly throughout the corona. He coined the term "nanoflares" to describe the myriads of very small reconnection events that could occur. Toshifumi Shimizu, a Japanese scientist, did find many tiny x-ray flares in Yohkoh satellite data, but they were too infrequent to supply the necessary power. Nanoflares have retained their appeal nevertheless. Eric Priest, Peter Sturrock, Jim Klimchuk, and many others have revisited the idea in some form. It seems that without an absolutely positive demonstration, no plausible mechanism is ever totally discarded.

Meanwhile, other researchers have turned to MHD waves as a way to heat the corona. Three types of MHD waves are important in plasma that is dense enough to behave like a fluid. The "slow mode" resembles sound waves but propagates solely along the lines of force at the speed of sound. In the fast mode, magnetic and plasma pressures combine to drive the wave; it can propagate across, as well as along, the field lines at speeds as large as or larger than the Alfvén speed. Motions in the wave are both along and transverse to the field direction.

Finally, the Alfvén mode propagates only in the direction of the field. It can distort field lines into three forms: a kink, like a snapped whip; a sausage, like the bulge in a balloon; and a twist, like a corkscrew. Alfvén waves do not compress or expand the plasma as they pass, which turns out to be both helpful and unhelpful.

Many authors have investigated how these modes might convey energy to the corona. Usually they just postulate that a flux of wave energy is available in a range of frequencies in the photosphere, without too much regard for the generation of the wave. Then they investigate how the wave propagates and dissipates. Often a theorist will try out an idea in the context of an active region loop, because it is simpler and more easily observed.

The slow mode has been thought to fail as a possible way to heat the corona because it shocks too low in the atmosphere. The fast mode also has problems in reaching the corona. In a diverging magnetic field and with plasma density decreasing with height, the fast mode turns back toward the surface (refracts) and does not penetrate to the corona to any great extent. It may, however, transfer its energy to other kinds of waves, as Donald Osterbrock of Lick Observatory suggested in an influential paper back in 1961. Thomas Bogdan and friends at the High Altitude Observatory are reexamining this process in numerical simulations.

But for many years the most promising MHD wave has been the Alfvén wave. It has received much attention over the past thirty years by a number of authors, but it has two serious problems. This mode can propagate easily into the corona along the field lines, but being noncompressive, it does not form shocks. Therefore, some other mechanism must be invoked to dissipate Alfvén wave energy. Second, the motions in the photosphere are slow, typically less than a kilometer a second. The amount of power that they could generate in the form of Alfvén waves is uncertain. In any case, Alfvén waves with the necessary amplitudes haven't been detected in the low atmosphere, despite intense searches.

Solutions have been proposed for both of these problems. Suppose, for example, that the waves are generated well below the photosphere. Numerical simulations of turbulent convection by Robert Stein and Äke Nordlund showed that a surprising amount of high-frequency sonic power is generated. In the presence of a strong magnetic field, some of the acoustic energy could be transformed into Alfvén waves. But it is only a suggestion.

Dissipating the energy of Alfvén waves is even more difficult. Viscosity, electrical resistance, and radiative losses are all too feeble as agents, except in the presence of strong gradients in the magnetic field strength. Joseph Davila at the Goddard Space Flight Center offered an important idea in 1987 on how such gradients might originate in coronal loops.

Davila imagined a magnetic loop, embedded in field-free plasma and standing with its legs planted in the photosphere. Suppose that subsurface motions excite Alfvén waves of different periods, say, between five and three hundred seconds. A wave whose period equals the time the wave requires to cross between the footpoints of the loop would be "resonant." It would be a standing wave and would be effectively trapped in the loop. This so-called body wave would set up surface waves in the thin interface with the surrounding field-free plasma. Davila showed how the surface waves would develop strong shearing motions that could dissipate the energy of the body wave uniformly along the loop, but whether the energy released at the loop's surface could penetrate into its body was debatable. Davila's model would have to wait several years for a test.

What Do Observers Tell Us?

How do these ideas about wave heating stack up against observations? Many past observers have reported oscillations of coronal brightness or velocity that they claimed as evidence for the presence of waves of some kind. Markus Aschwanden, at the Lockheed Martin Solar Group, recently summarized these observations.

Markus is a remarkably hard-working scientist. He began his career as a radio astronomer in Switzerland, where he carried out research on the plasmas of solar flares. Eventually he migrated to the University of Maryland, next to the NASA Goddard Space Flight Center and then west via Stanford University to join Alan Title's group in Palo Alto, California. He is one of those rare individuals who possesses talents as an observer, analyst, and theorist.

In his review of coronal oscillations, he found examples from all over the spectrum, from x-rays to radio waves. Brightness and velocity variations have been detected, but seldom simultaneously. Periods have been observed as short as a few seconds at radio wavelengths and as long as thirty minutes in coronal spectral lines. The oscillating structures were usually unresolved, which is certainly a limitation, but that did not hamper attempts to interpret the data as some form of wave.

I have to say that many of these identifications seem speculative to me. Without corroborating evidence, such as combined velocity, brightness, and density fluctuations, it is difficult to choose a particular type of wave with any certainty.

However, in 1998, Aschwanden and his co-workers made an unambiguous identification of an Alfvén wave. For the first time they observed wiggling motions along a loop that clearly indicate the presence of coronal Alfvén waves. They benefited from the extremely good spatial resolution (a few hundred kilometers) of TRACE images that allowed them to see definite lateral movements of the main body of the loop. The images were recorded in two spectral lines, at 17 and 20 nm, that are emitted at 1 and 2 to 5 million degrees, respectively.

After a small flare, they saw five loops in an active region oscillating *perpendicular* to the long axis of the loop. The oscillations persisted for three to five periods, with an average period of 280 seconds. The researchers decided that the kink Alfvén mode was the most likely.

These beautiful observations show that a shock wave from a flare can excite Alfvén waves in active region loops. From such data Aschwanden hopes to be able to derive the strength of coronal magnetic fields and other properties of coronal loops. But these results say nothing about the role of Alfvén waves in heating the *preflare* loops. These waves are apparently much weaker and have gone undetected so far.

In later work, Aschwanden found other examples of flare-excited Alfvén waves, using data from TRACE. He detected kink mode oscillations that died out after two or three periods. He therefore had an opportunity to test several dissipative mechanisms, including Davila's resonant absorption idea. To make a long story short, he was able to rule out all but the resonant absorption mechanism. In principle it should be possible to determine the rate of energy input from the Alfvén waves, but this hasn't been done yet.

Another way to identify the mechanisms that heat coronal loops is to ask where the heat is deposited and to compare the results with predictions of different models. So far this route has not proven helpful; different researchers with different data arrive at different conclusions.

For instance, Robert Rosner and friends at Harvard University learned from their pioneering study in 1978 that a static loop must be heated at its *top*. Using this constraint, they devised a universal scaling law that relates the plasma pressure, temperature, and loop length in many different loops. They concluded that Alfvén waves and DC current heating were compatible with their observations but that acoustic heating was not.

In 2002, Eric Priest and co-workers suggested a reconnection scenario that would produce *uniform* heating. They drew attention to the intense small-scale magnetic fields that Alan Title and his group had discovered. These tiny loops are jammed into the borders of the supergranule cells and are constantly colliding. Priest suggested that continual reconnections among them would release energy. Parts of the overlying corona would be heated *uniformly,* although most of the released energy would be dissipated in the chromosphere.

Markus Aschwanden has arrived at a completely different picture. He reviewed the whole subject of coronal heating in 2001, relying on recent results from TRACE, SOHO, and Yohkoh. He listed three observational constraints that a theory of heating should explain. First, the heat input is confined to the footpoints of a loop. (The source of heating, however, was left open.) Second, a loop is not in static equilibrium; jets of hot chromospheric plasma flow up its legs and raise their density beyond what gravity alone would allow. Finally, the jets convey energy to higher parts of the loop. So in his view, waves or currents may play a secondary role in heating the upper parts of a loop. Instead hot jets may carry the energy.

Aschwanden concluded that neither models of MHD wave heating nor reconnection have developed far enough for a quantitative comparison with the observations at the footpoints of loops. All candidate mechanisms, including acoustic heating, are still on the table.

This state of affairs may seem either disappointing or refreshing, depending on your point of view. We now have masses of observations of the corona as well as many viable suggestions for heating mechanisms. Perhaps the corona is heated in a variety of ways and everybody is correct. Perhaps we have all missed the essential factor. But we will not know until theory catches up with observation, to the point where critical comparisons can be made and some ideas are put to rest.

The coronal heating problem extends out into the solar wind, where new factors come into play. We'll turn to these next. We'll begin with a story about comets.

A HOT WIND FROM THE SUN

Many people have never seen a comet. Naked-eye comets are relatively rare, and with urban lights and country clouds, they are easy to miss. But if you've ever had the pleasure of seeing a comet, you know they have long feathery tails.

In 1950 Ludwig Biermann noticed something odd about comet tails. He realized that a tail always points away from the Sun, regardless of where the comet is in its orbit. From spectroscopic studies he knew that the tail consists of gases like carbon monoxide that have warmed up in sunlight and blown off from the comet's icy nucleus. Therefore, he might have expected that the tail would trail directly behind the nucleus as the comet wends its way around the Sun. But that's not what he saw.

To explain this curious effect, Biermann postulated that the Sun emits an intermittent rain of "corpuscles," which pushes on the comet tails and points them away from the Sun. Because comets show this effect wherever they are in the sky, Biermann concluded, the Sun emits these corpuscles in all directions. From observations of the acceleration of blobs in the tails, he could estimate the speed of the particles as 500 to 1,500 km/s and their density at the Earth's orbit as at least $500/cm^3$.

Eugene Parker at the University of Chicago followed up on Biermann's findings. In a seminal paper of 1958, Parker outlined a theory that the hot corona of the Sun would naturally expand into space as a "solar wind." We know that some unknown process maintains the corona at a temperature exceeding a million degrees, he wrote. At such temperatures, the coronal gases are fully ionized. The thermal pressure of the coronal plasma causes it to expand, but the Sun's gravity tends to restrain it. Parker showed that gravity would modify the expansion but that beyond a critical radius where the speed of sound is reached, a supersonic flow would escape the Sun. In effect the thermal energy of the corona was converted to kinetic energy, as in the escape of air from a tire (note 5.2).

Parker also showed how the Sun's dipole magnetic field, which was thought to extend from pole to pole, would be dragged out into space by the wind. Because the Sun rotates as the wind expands radially, the field lines would be curved into a flat Archimedes spiral, at least near the equator (note 5.3).

Parker's theory was received skeptically and debated for several years. Then in August 1962, NASA launched Mariner 2 to survey the atmosphere of Venus. On its three-month voyage to Venus, instruments aboard the satellite confirmed Parker's prediction of a solar wind streaming radially outward from the Sun. Parker's theory was hailed as a stunning achievement, which has since opened a vast field of research.

The wind was not smooth, though. Conway Snyder and Marcia Neugebauer, from the Mariner team at NASA's Jet Propulsion Laboratory (at Caltech), reported in 1962 that the wind is composed of discrete streams with speeds as high as 700 km/s and as low as 400 km/s. The streams rotate with the Sun, and several of them recurred in successive rotations. The reappearance of these streams coincided with strong disturbances (storms) in the Earth's magnetic field.

Then in November 1963, the Interplanetary Monitoring Platform (IMP 1) went into near-Earth orbit to investigate the Earth's extended magnetic field. The magnetometers aboard it revealed that the interplanetary magnetic field (IMF) near Earth was arranged in two to four unipolar "sectors." The sectors rotate with the Sun, and as each one sweeps by the Earth, its field strength varies from a maximum of about 60 to 40 microgauss.

NASA scientist Norman Ness and John Wilcox at the University of California at Berkeley took the next step. They reported in 1965 that they could relate the magnetic polarity of a sector to a source area in the photosphere, if they took into account Parker's prediction of spiral field lines. They could follow a field line back to a specific area of the photosphere. Their result confirmed another of Parker's predictions, the Archimedes spiral. But there was nothing special in the target areas in the photosphere that would suggest an origin for the sectors or the streams.

In table 5.1, I've listed some of the properties of the fast and slow wind streams near Earth, as observed near the minimum of the solar cycle. The most obvious difference is the wind speed, as low as 250 km/s in slow streams and as high as 800 km/s in fast streams. The proton density is higher and the proton temperature is much lower in slow streams. Notice also that the temperatures of electrons and protons differ in each type of stream. This result arises from the very low plasma densities, in which the particles collide so infrequently that they can acquire independent temperatures. Indeed, in this "collisionless" plasma, particles interact through the action of a variety of waves.

So by 1963 we knew that the Sun blows a blustery wind into space. Looking back in time, this result should not have surprised us. Geophysicists had known since 1904 that the Sun probably emits narrow streams of "corpuscles" that shake the Earth's extended magnetic field. In that year E. Walter Maunder, a solar astronomer at the Royal Greenwich Observatory, had demonstrated that the greatest of these geomagnetic storms recur with a period of twenty-seven days. This is the Sun's synodic rotation period, which is the time for a fixed feature on the Sun to rotate to the same apparent position as viewed from Earth.

Table 5.1. Properties of the average fast and slow wind near the Earth

Property	Slow wind	Fast wind
Speed (km/s)	350	700
Plasma density (cm^{-3})	12	4
Temperature (K)		
Protons	34,000	200,000
Electrons	130,000	100,000

Maunder postulated that somewhere on the Sun were long-lived sources of corpuscular streams. The streams rotated with the Sun, swept past the Earth (rather like a lawn sprinkler), and caused the storms. From the duration of the storms, he could estimate the width of the streams, about 20 degrees in longitude. Many streams were associated with the appearance of large groups of sunspots on the solar disk, but some streams persisted long after the sunspots disappeared or appeared before the sunspots erupted. He concluded that sunspots were not necessary for a stream's existence. Some unknown areas on the Sun were responsible. They were named M regions by Göttingen professor Julius Bartels in 1932.

Many attempts were made later to identify M regions with known objects, such as active regions or filaments, but they all failed. The nature of M regions remained a mystery until 1972. Three independent lines of research came together in that year to provide a positive identification.

Where the Wind Rises

The first of these three was a Harvard College experiment aboard Orbiting Solar Observatory 4 that obtained images of the low corona in 1967 at wavelengths between 30 and 140 nm. The images showed regions ten times dimmer than anywhere else. They were named "coronal holes." They were characterized by temperatures 600,000 degrees cooler, and plasma densities three times lower, than normal coronal regions.

Next, several scientists concluded that the fast streams observed in the wind could only escape regions of the corona where the magnetic field opened outward to space, not from the strong closed fields of active regions. The coronal field is not observable, but Martin Altschuler at the High Altitude Observatory showed how to calculate a potential field, starting from observations of the photospheric field that covered the entire Sun. When he and his colleagues compared his coarse reconstructions of the global field (called "hairy balls") with coronagraph observations of coronal holes, he found that the fields in the holes did diverge and open outward. They were prime candidates for the locations where fast streams might originate.

Allen Krieger and his team at the American Science and Engineering Corporation, in Cambridge, Massachusetts, completed the chain of evidence connecting high-speed streams to coronal holes. Krieger and friends were in the business of designing x-ray scanners for airports, but their real love was solar physics. They had developed a novel telescope capable of forming sharp images in soft x-rays (3 to 5 nm) and launched it several times on short flights of a sounding rocket. Krieger's group was one of a number around the world that was opening the skies to x-ray astronomy.

On November 24, 1970, Krieger obtained a crucial x-ray image of the solar corona. Active regions were bright and filled with loops; most of the disk was fainter and diffuse; and a dark coronal hole extended from the equator to the south pole. Faint x-ray threads appeared to diverge from the hole. Altschuler provided a coronal magnetic field reconstruction for the day, which showed that the hole did have diverging field lines.

Then in 1973, Krieger and his team used published satellite observations of interplanetary wind speeds to trace a recurrent fast stream back to the Sun using the spiral field geometry. The stream mapped back into the hole that they had observed in 1970.

Krieger and his team were ecstatic, and with good reason. They had demonstrated for the first time that a coronal hole was the source of a fast wind stream. Because such streams were known to produce geomagnetic storms, they could conclude that they had found Bartels's mysterious M regions. One example doesn't prove a case, it's true, but the Skylab satellite was soon able to confirm and extend their result.

As a footnote to the discovery, the team noted that the temperature of the hole, as determined from the brightness of the soft x-rays, was only 1.3 million degrees. The photospheric field was also weak in the hole. Yet the stream from the hole had a speed of almost 600 km/s. This result was a dark cloud looming over Parker's theory, as we shall see later.

The theorists were also making progress. In 1971, Roger Kopp and Jerry Pneuman at the High Altitude Observatory published a dynamical model of the heliosphere that proved to be prophetic. They recognized from eclipse photographs that the solar wind must flow around the closed field lines at the base of a coronal streamer. If all the streamers were located in an equatorial belt, as happened around solar minimum, the polar caps would be covered with open field lines. Moreover, these polar lines of force would meet at a neutral sheet that extends outward from the streamer belt.

They used this concept to set up a model of the global field at solar minimum. Assuming an isothermal corona, they then calculated the equilibrium wind velocity and field strength that would prevail all around the Sun. The balance of gas and field pressures, gravity, and inertia determined the shape of the field and the streamlines. Figure 5.2 shows their result. Both poles of the Sun are in effect huge coronal holes and the sources of the wind. We'll refer to this figure later in connection with the slow solar wind.

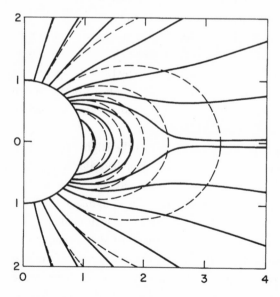

Fig. 5.2. In this model of the global magnetic field around sunspot minimum, the field lines are pushed down toward the solar equator by the expanding wind from the polar coronal holes. The dashed curve shows the dipole field that would prevail without a wind.

The Shape of the Wind

The Skylab mission proved beyond all doubt that coronal holes are the sources of high-speed wind streams. It also provided the data for a new concept of the overall shape of the interplanetary field and the wind. The new concept would be confirmed by the flight of Ulysses over the poles of the Sun, two decades later.

Skylab orbited the Earth from May 1973 to February 1974, during the decline of the sunspot cycle. The Sun was very quiet during this period, and the corona had a particularly simple shape. Both of these factors helped the scientists to interpret their data.

Skylab was equipped with a battery of x-ray telescopes to image the corona on the disk and a white-light coronagraph to record it beyond the edge of the Sun. Many ground-based observatories supplemented Skylab's instruments. In particular, the coronagraph at the Mauna Loa Observatory in Hawaii monitored the brightness of coronal streamers throughout the flight. A three-dimensional model constructed from these data showed that a nearly continuous belt of helmet streamers girdled the Sun near its equator like the brim of a battered sombrero. The "brim" changed shape during the mission, tilting north or south at some longitudes by as much as 30 degrees. As we shall see, this change of shape would prove to be an important clue.

The x-ray telescopes revealed that the poles of the Sun were huge coronal holes, with opposite magnetic polarities. Whenever a polar hole rotated onto the visible solar disk, a stream was detected at Earth by the IMP spacecraft, with speeds between 400 and 750 km/s. The magnetic polarity of the stream matched that of the hole. Here was incontrovertible evidence that high-speed streams do in fact originate in coronal holes.

The best was yet to come. Arthur Hundhausen at the High Altitude Observatory noticed two occasions when each polar hole had dipped down to cover the Sun's equator for a full 180 degrees of longitude. Two enormous wind streams were observed then, each with the polarity of one of the poles and with wind speeds of 750 to 800 km/s. Two magnetic sectors of opposite polarity dominated the interplanetary field for many rotations thereafter.

Hundhausen deduced from these events that the global field of the Sun had the shape of a tilted dipole. He pictured the magnetic equator of the dipole as coinciding with the streamer belt, tilted some 30 degrees from the equator. As the dipole rotated about the Sun's axis, the IMP satellite detected a two-sector interplanetary field and two powerful wind streams. His main point was that the wind we observe at Earth originates in the polar caps of the Sun and bends down into the orbital plane of the Earth (the ecliptic).

We should emphasize that the IMP satellites, like all others around this period, were limited to orbiting in the plane of the ecliptic. Because they could sample the solar wind only in that plane, there was no direct way to confirm Hundhausen's concept of a dominant polar wind during the minimum of the solar cycle (note 5.4).

With the launch of the Ulysses satellite in October 1990, however, a new era dawned in the study of the heliosphere. This spacecraft was the first to explore the heliosphere within 5 astronomical units (AU) of the Sun and over all latitudes. To reach the Sun at high latitudes, the craft had first to journey to Jupiter in order to pick up a gravitational boost. The trip took four years and was itself a valuable survey of interplanetary space.

The main objective of the mission was to fly by the poles of the Sun. In 1994–1995 the satellite passed over the Sun's poles at distances between 1 and 5 AU, sampling all the properties of the wind and magnetic field. These passes occurred during the decline of the solar cycle, a phase similar to that of the Skylab era.

Ulysses confirmed many of the features of Hundhausen's tilted dipole model. Each polar cap emitted a thin steady wind with speeds as high as 750 km/s. These polar winds expanded rapidly in each hemisphere, filling most of the volume around the Sun and dipping down to the ecliptic. A slow dense wind of 400–500 km/s centered on the magnetic equator separated these polar flows.

The magnetic lines of force coincided everywhere with the streamlines of the flow, just as in the model of Kopp and Pneuman. As in their model, the flow near the surface of the poles was strong enough to change the direction of the polar fields. Instead of expanding radially, the streamlines curved outward sharply. Farther from the poles the wind was flowing radially. As the Sun rotated, the field was wound into an Archimedes spiral. Figure 5.3 shows the complex form of the interplanetary field in 1995.

Ulysses passed over the poles for a second time in 2000 and 2001. This time the solar cycle was near maximum and the global structure of both the

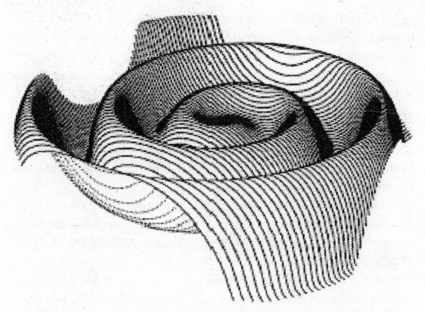

Fig. 5.3. Near the sunspot minimum the global field of the Sun has the shape of a dipole whose axis is tipped with respect to the Sun's rotational axis. The wind from the polar holes drags the field lines down to the "neutral sheet" at the equator of the dipole (as in fig. 5.2). The field points outward above the sheet, inward below the sheet. As the Sun rotates, the field lines at the neutral sheet form this wavy Archimedes spiral in interplanetary space.

surface magnetic field and the wind were far more complicated. The polar holes had shrunk dramatically, and slow wind was observed at all latitudes. Observations from SOHO showed many small holes at all latitudes, but the correspondence between a hole and wind streams was weak.

By 2002, the polar holes were growing equatorward again and the fast polar wind was restored. Evidently uniform fast winds prevail for most of the solar cycle, with only a one- or two-year break. This short period coincides with the reversal of the polarity of the polar fields.

As of this writing (December 2007), Ulysses is passing over the Sun's north pole for the third time. Once again the solar cycle is near a minimum, so we should expect to find the heliosphere in much the same condition as in 1994–1995. But wait and see!

What's under the Hood? The Engine That Drives a Stream

Since the discovery of the solar wind, theorists have struggled to explain how fast and slow streams are accelerated. With fast streams, they start with several advantages. First, they know that the sources of the fast streams are in coronal holes. There is no general agreement yet on the sources of the slow wind, although helmet streamers are considered the best candidates. Second, fast streams are steady and fairly uniform in properties, whereas slow streams are highly variable. For these reasons and others, fast streams have received most of the attention of modelers and theorists until recently.

In what follows, I'll focus on just a few of the many researchers working on this subject. It would take us too long and too far afield to try to cover the work of even a minority of the scientists active in this blossoming discipline.

The basic problem of the fast streams is accounting for the amount of energy they carry. In Parker's hydrodynamic theory, the total energy of a stream (thermal plus kinetic plus gravitational) derives from the amount of heat injected at the base of the corona. To match the speeds observed at Earth, Parker required a base temperature of at least 2 million degrees, which was quite reasonable at the time. But several lines of evidence have indicated since then that temperatures are no higher than 1.6 million and possibly as low as 0.9 million degrees in coronal holes.

Parker actually realized the limitations of his thermal model as early as 1965, years before coronal holes were identified. His model predicted wind temperatures that were much too low near the Earth in comparison with observations. He concluded that additional energy of some type must be deposited high above the base of the corona.

In the 1980s, Egil Leer and Tor Flå (University of Tromsø, Norway) and Thomas Holzer (High Altitude Observatory) carried out a series of calculations designed to pin down where the additional energy was deposited. They concluded that most of the energy must be injected above the height at which the wind speed reaches the sound speed (or in the presence of a magnetic field, the Alfvén speed). This was essential to further increase the wind speed, because energy addition in the subsonic region only increases the flux of mass rather than the speed.

Researchers also recognized that the additional energy would have to be nonthermal, and most likely some form of MHD wave. As we noted earlier, the fast mode and the slow mode each has its drawbacks as a means of heating the corona. The most attractive candidate was an Alfvén wave that is generated near the Sun's surface and propagates outward, depositing energy and momentum as it decays.

Weak Alfvén waves were actually observed in the wind near the Earth's orbit in 1971. Joe Hollweg, a perennially youthful professor at the University of New Hampshire, was one of the first to follow up on this discovery. He has been an active proponent of Alfvén wave heating and acceleration ever since. As early as 1973 he postulated that the Alfvén waves observed at Earth were the remnants of a powerful flux leaving the Sun.

He showed how such waves could accelerate the plasma to high speed by a combination of wave pressure (note 5.5) and wave heating. His theory also predicted the higher temperature of protons relative to electrons that was observed near Earth. But his theory failed to account for the rapid and continuous acceleration observed within the first few radii of the Sun. The fault seemed to lie in the mechanism Hollweg adopted for dissipating the waves.

As we mentioned earlier, kink mode Alfvén waves don't compress the plasma as they propagate. Therefore, they don't form shocks in the plasma, which could dissipate wave energy. Some other means of dissipating an Alfvén wave must be identified if they are to be useful as heating agents. Much of the current research on fast wind streams is aimed at investigating possible processes.

In the 1980s, experiments aboard the Helios solar probe obtained new data at about half the distance to the Sun that pointed to an attractive process. Helios discovered that the hot protons in the fast wind have more thermal energy (i.e., faster random motions) in the direction *perpendicular* to the interplanetary magnetic field than in the longitudinal direction. Another way of saying that is that the proton temperature transverse to the field is higher than in the parallel direction.

This was clear evidence that collisions between particles are too rare to be important in redistributing energy. The plasma doesn't act as a smooth, continuous fluid and must be treated as an assembly of particles. In such collisionless plasma, say, beyond 3 solar radii, waves of various types serve to transfer energy between particles (note 5.6). To explain the two different proton temperatures, several researchers proposed a special type of wave, the so-called ion-cyclotron wave.

Let's take a moment to consider such waves. First, recall that ions in magnetized plasma gyrate around the lines of force because of the Lorentz force (see chapter 1). A positive ion, like a proton, gyrates counterclockwise around a field line that is directed outward from the Sun. Its frequency of gyration depends on the field strength and the ion's mass-to-charge ratio. Now if an Alfvén wave with the same frequency and with a left-handed twist passes by, the wave will couple to the ion and accelerate it. The ion will gain energy in the direction of gyration, or *transverse* to the field line. So an ion-cyclotron wave has exactly the properties needed to account for the higher transverse temperatures observed in protons.

So far, so good. The problem arises that wave frequencies of a few *kilohertz* are required to match the proton gyration frequencies. No waves like these have ever been observed directly. Moreover, if Alfvén waves were generated in the photosphere, they would have frequencies typical of the granular motions there, in the range of 0.01 to 1.0 hertz. How could such low-frequency waves transform into the required high-frequency waves?

In 1988, Joe Hollweg and co-author Walter Johnson conceived a different way to generate high-frequency waves: in the wind, rather than in the photosphere. They proposed a model in which low-frequency Alfvén waves, propagating from the Sun's surface, break down into waves of higher frequency in a so-called turbulent cascade. Such a cascade occurs, for example, when the water flowing in a creek encounters a rock. The smooth flow breaks down into eddies of all sizes behind the rock. The smallest eddies dissipate energy as heat.

Hollweg and Johnson showed that, with reasonable assumptions, a small portion of the high-frequency waves could be available to heat protons by means of the ion-cyclotron resonance. In their model the pressures of hot protons and waves combine to drive the wind to high speed below 3 solar radii, in good agreement with observations.

Ian Axford and his colleague J. F. McKenzie at the Max Planck Institute for Aeronomy in Lindau, Germany, were skeptical about the turbulent cascade scenario. In 1992 they proposed instead that the strong, mixed-polarity magnetic fields observed at the boundaries of supergranule cells might annihilate in "microflares." (This was actually observed later on.) These tiny explosions might then generate the high-frequency hydromagnetic waves needed to heat the fast streams and incidentally to heat selected ions by ion-cyclotron resonance.

Their proposal was explored by Eckardt Marsch and C.-Y. Tu, also at the Max Planck Institute for Aeronomy. In 1997 they postulated a spectrum of waves with frequencies between 1 and 800 hertz and with enough power to heat the wind. They followed the dissipation of the waves by the ion-cyclotron resonance and showed that coronal temperatures and a respectable wind speed could be reached within a few solar radii. Some critics were not convinced by their demonstration, though, because of the strong assumptions they were forced to make.

Further evidence that the ion-cyclotron resonance could be crucial in heating the wind appeared after the launch of the SOHO satellite. In 1998, John Kohl and his associates at Harvard University obtained ultraviolet spectra of highly ionized oxygen atoms in a polar coronal hole, out to 4 solar radii. The profiles of the spectral lines were extremely wide, indicating transverse temperatures as high as *200 million* kelvin. The temperature along the magnetic field was ten to one hundred times smaller. Similar results were obtained for an ion of magnesium. Here was a smoking gun pointing to ion-cyclotron resonance.

However, oxygen and magnesium are relatively rare elements in the wind plasma. How could their ions act to heat the great bulk of the plasma in the wind? Steven Cranmer, a young astrophysicist in Kohl's group, took the first step toward answering this question by devising a quantitative theory of ion heating by the cyclotron resonance.

He first assumed that a flux of high-frequency waves exists in the wind and then showed how they could produce the observed transverse temperatures of ionized oxygen ions. His theory didn't do as well in predicting the flow speeds of these ions along the field. But such ions could heat protons by collisions, thus leading to the possible acceleration of the great mass of the plasma.

Cranmer learned that the heating process is so efficient that the flux of high-frequency waves is strongly attenuated within a short distance from the Sun. He concluded that above 1.5 solar radii, processes within the wind must replenish the waves. One possible source was Hollweg's turbulent cascade of low-frequency Alfvén waves.

So in a following paper, in 2003, Cranmer and his colleague Adriaan Van Ballegooijen reexamined Alfvénic turbulence and proton heating. Van Ballegooijen (a Dutch name pronounced "Balla-goi-hen") has worked on some of the most difficult problems in coronal physics and the solar cycle, with considerable success. The two men make a formidable team.

The details of their theory are too technical to relate here. Suffice to say that they modeled the generation of high-frequency waves by the breakdown of low-frequency waves in a turbulent cascade. In turbulence we are not dealing with coherent trains of waves but random fluctuations of magnetic field and velocity that momentarily show the characteristics of an Alfvén wave. Waves whose frequency matches the proton's gyrofrequency and which travel along the field lines in either direction can heat the protons by the cyclotron resonance.

As a last step, the two researchers estimated the amount of high-frequency power available to heat protons and heavier ions by the cyclotron resonance. They learned, to their considerable disappointment, that there is insufficient power. The ion-cyclotron scheme was in trouble.

In 2005, Cranmer and Van Ballegooijen decided to back off from the alluring ion-cyclotron scenario and examine, as realistically as possible, how Alfvén waves are generated, how they propagate, and whether they really have sufficient energy to accelerate the wind. Only then would they adopt a particular heating mechanism to investigate whether Alfvén waves would deposit their energy at the appropriate distances from the Sun.

Unlike all other modelers they derived the initial spectrum of possible Alfvén waves by examining the motion of so-called magnetic bright points (MBP). These are regions of kilogauss magnetic fields, no larger than about 300 km, that occur in the dark lanes between solar granules. Richard Muller at the Pic du Midi Observatory had observed the displacements and velocities of MBP. Cranmer and Van Ballegooijen performed a Fourier analysis on Muller's data and found periods ranging between 0.3 and 2,000 minutes. These are the possible initial periods of Alfvén waves at the solar surface. As expected, most of the energy is concentrated at periods of about 1 to 10 minutes.

Next they adopted a complete model of the magnetic field, plasma density, and velocity in a spreading magnetic tube that extends up to 4 solar radii from the boundary of a supergranulation cell. The small loops near the photosphere merge at some height into a uniform rapidly spreading flux tube.

With this tube model and with the initial Alfvén spectrum, they were able to calculate how waves of different periods propagate and are reflected. They discovered that only about 5 percent of the waves reach the corona and wind directly. The rest is reflected and interacts with arriving waves, a process that is important in generating waves (or magnetic fluctuations) of much higher frequencies.

As the Alfvén waves travel out into lower plasma densities, their velocity amplitudes increase. If no dissipation mechanism existed, their amplitudes would combine to reach a maximum of 400 km/s at distances of 1 to 100 solar radii. That would be ten times too large to match radio observations of turbulent velocities in the wind.

But now the authors invoke a turbulent cascade mechanism, in which inward and outward traveling waves collide and feed energy to higher and higher frequencies. They didn't mention the ion-cyclotron resonance, though; the turbulence is just assumed to end in fluctuations small enough for viscosity to convert them into heat.

Indeed, their calculations predicted extended heating of the plasma between 1 and about 20 solar radii, with velocity amplitudes in much better agreement with observations. Wave pressure was also important at this range of distances. The acceleration due to waves exceeded that of gravity at about 1 solar radius, just where the wind begins.

As the authors freely acknowledged, their calculations of wave heating and pressure would not necessarily maintain the steady-state model they started with. Ideally one would want the initial Alfvén flux to distribute energy at just the right places to accelerate the wind in a flux tube whose shape and plasma density are, in turn, determined by the deposition of energy. We shall see later on how these authors succeeded in building a self-consistent model for the solar wind.

The Slow Wind

We turn now to the slow wind, with speeds between, say, 250 and 400 km/s. We saw in table 5.1 how the slow and fast streams differ in most of their basic properties. The task of the theorist is to explain these differences in physical terms. The slow wind is intrinsically more difficult to understand, however, first because it is unsteady and highly structured; and second because the exact locations of its sources are still uncertain and seem to vary with the solar cycle. As a result, there is no consensus on many of the issues involved in the slow wind. Several viable proposals for the source and acceleration of the wind are still on the table. Models of heating and acceleration are almost completely lacking for the slow wind, in contrast to the many mature models proposed for the fast wind.

Ulysses observations showed clearly that, around solar minimum, the slow wind originates from the streamer belt that hugs the solar equator (fig. 5.4, a polar speed plot). That is well and good, but researchers still disagree on whether slow wind emerges from the tip of a streamer or from a sheath that encloses the streamer or from the low-latitude boundary of the huge polar coronal holes. There are good arguments on all sides, as we shall see.

One key factor relevant to all of these schemes is the rate at which open field lines spread apart in the corona. In 1990, Y.-M. Wang and Neil Sheeley at the U.S. Naval Research Laboratory discovered that the speed of a stream near Earth depends inversely on this rate. Thus pipe-like flux tubes in the corona yield high-speed streams, as at the centers of large coronal holes, and trumpet-shaped tubes yield slow streams. This concept was later extended to the rate of field line spreading around streamers and at the borders of coronal holes.

Figure 5.1 shows several streamers in white light as observed by the LASCO coronagraph on April 27, 1998. The accompanying drawing illustrates some of the key parts of a streamer: the "helmet" of closed field lines; the "cusp," where open and closed field lines meet at a neutral point; and the narrow "stalk" along the axis. Notice also the faint streaks of emission that surround the brighter helmet and stalk. Wind of some type is evidently flowing out in this zone.

In 1997, Neil Sheeley and a huge list of co-authors reported on flow speeds from the streamer belt. They examined movies of streamers, obtained near solar minimum by the LASCO coronagraph and discovered that discrete blobs of plasma are continuously squirted out of the tops of streamers. The blobs originate at about 3 to 4 radii from Sun center, above the cusps of streamers. They typically reach a speed of 150 km/s at 5 radii and accelerate to 300 km/s at 25 solar radii. Although the blobs are still accelerating at 25 radii, these data do suggest that the slow wind originates in the *stalks* of streamers. The existence of blobs could help to explain why the slow wind is observed to flow so erratically.

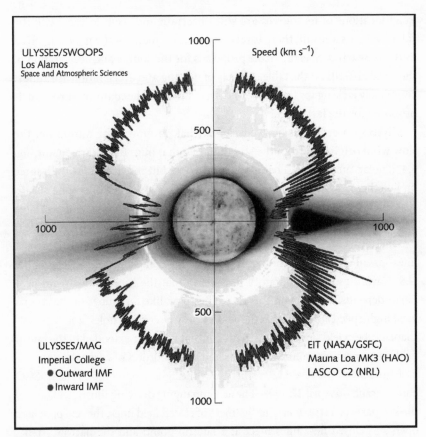

Fig. 5.4. This diagram shows observations of solar wind speeds obtained by the Ulysses space probe after circling the poles of the Sun. At any angle around the Sun, the length of the radius is proportional to the speed. During this period around sunspot minimum, both northern and southern hemispheres were filled with fast wind (750 km/s), while the equatorial belt emitted slow wind (around 400 km/s).

Shadia Habbal at the Harvard Smithsonian Observatory, Richard Woo at the Jet Propulsion Laboratory, and their colleagues agree that the stalk is the source of the slow wind. From the Ultraviolet Coronagraph Spectrometer (UVCS) spectra of highly ionized oxygen (O^{5+}), LASCO images, and radio observations of several streamers, they determined a typical speed of 94 km/s in the stalk at 4.5 solar radii. Moreover, they detected a sharp transition to 180 km/s outside the stalk, with indications of a faster and hotter wind. So we arrive at a picture of slow cool wind emerging from the stalk and faster wind flowing past the streamer.

Steven Suess of the NASA Marshall Space Flight Center and Steven Nerney of Ohio University took a contrarian view, however. In 2004 and 2005, they pointed out that the field lines outside the streamer are likely to diverge rapidly with height, as in the model of Pneuman and Kopp (fig. 5.2). *Here,* they claimed was the more likely source of the slow wind, not primarily in the stalk. Suess and Nerney computed analytic models to support their concept, but as far as I know, no observations have been made to either confirm or disprove their assertion.

Slow Wind during the Solar Maximum

The sources of slow wind during the solar maximum are far more controversial than during the solar minimum. During solar maximum, the corona is a complicated mixture of active region loops, small coronal holes, and active region streamers. They are all candidates as sources of the slow wind. Observations from Ulysses and SOHO are not particularly helpful in sorting out the possibilities. They present a very confusing mixture of fast and slow streams that seem to originate at all solar latitudes and are highly variable in speed and lifetime. So it is not surprising that the experts differ.

In 1997 S. Bravo and G. A. Stewart at the National Autonomous University of Mexico proposed the borders of coronal holes as sources. They pointed to potential field reconstructions of coronal holes that show strongly diverging field lines at their borders. The famous hydromagnetic model of Pneuman and Kopp (fig. 5.2) also shows rapidly diverging field lines at the poles. And as we saw earlier, the rate at which field lines diverge is a good indicator of the speed of the wind that escapes along those lines. Bravo and Stewart backed their proposal with a simple thermal model, to show they could match the slow speeds observed by Ulysses.

Richard Woo and Shadia Habbal, on the other hand, pointed to *active regions* as a source of slow wind. They observed the corona above an active region on the edge of the Sun, using the UVCS instrument as before. The ratio of the intensities of the two oxygen (O^{5+}) lines (at 103.2 and 103.8 nm) yielded flow velocities in the range of 100 km/s out to 5 solar radii. Here was hard evidence that slow wind can originate in or around an active region.

We tend to think of active regions as composed entirely of closed loops, but reconstructions of coronal magnetic fields do, in fact, show open field lines at their borders.

Woo and Habbal buttressed their proposal with an independent result from Ulysses. Observations with the Solar Wind Ion Composition Spectrometer confirmed that slow wind possesses a peculiar chemical composition. Elements such as magnesium and iron, with first ionization potentials of less than 10 volts, are more abundant by factors of three or more than they are in the photosphere. Such enhancements (called the first ionization effect, or FIP) are also seen in the corona of active regions. In contrast, fast winds have no FIP effect.

The reasons for the enhancements are controversial. Trapping of plasma in closed field lines seems to be essential to the effect. Special diffusion or acceleration processes are also thought to be involved. But in any case, the FIP effect does seem to point to active regions as possible sources of slow wind, as Woo and Habbal have argued.

Leonard Fisk and Nathan Schwadron at the University of Michigan advanced yet another scenario. They hypothesized in 2001 that the total flux of open magnetic field *remains constant over the solar cycle*. They reasoned that open field lines could only be destroyed if opposite polarities reconnect, and that is unlikely because open field occurs in large unipolar regions.

They described a type of diffusion of field lines that occurs by reconnection with the small, randomly oriented magnetic loops that are known to cover the solar photosphere. Their theory predicts that the diffusion causes field lines to cluster into coronal holes; to reverse the polar fields during the solar cycle; and to feed mass and energy into the solar wind. Moreover, the theory yields a simple inverse relationship between the coronal temperature of a source of wind and the wind speed.

That's quite a lot for a theory to predict! And indeed their work was greeted with some skepticism until observations from Ulysses confirmed in 2001 that Fisk's predicted temperature–wind speed formula was satisfied by both fast and slow winds (see table 5.1). That result doesn't necessarily validate the rest of their theory, but it, plus the FIP effect, does support the conclusion that active regions can and do produce slow wind.

Acceleration Models for the Slow Wind

Several enterprising theorists have attempted to construct full-scale models for the acceleration of the wind by waves. Foremost among these is Leon Ofman of the NASA Goddard Space Flight Center. Ofman is a specialist in plasma physics who is well known for his elaborate models of the solar wind. He has been especially active in applying multifluid descriptions for the collisionless wind plasma.

In 2000, Ofman offered a model for the slow solar wind from a streamer. He constructed equations to track the temperature and flow of three types of particles: protons, electrons, and helium nuclei (alpha particles). He included electron heating by electric currents; proton heating by viscosity; and thermal heat conduction along the field lines. Most of the heating arises from the Alfvén wave flux he adopts, with frequencies between 3 and 300 hertz. With these ingredients Ofman was able to reproduce the speeds observed with the UVCS that we described earlier. It's difficult to assess just which factors were most important in achieving Ofman's nice results, though, because so many are in action.

I'll end this chapter by mentioning a set of self-consistent models of the solar wind published in 2007 by Steven Cranmer, Adriaan Van Ballegooijen, and Richard Edgar. Their models are less realistic than Ofman's in that they treat the plasma as a single fluid, not as a collection of particles. Moreover, their models are only two-dimensional. However, their work has two great strengths. First, it relies on an Alfvén flux and an acoustic flux as derived from actual motions at the chromosphere, not some ad hoc assumption; and second, it includes the chromosphere, corona, and wind in one unified model. They demonstrate once again that the essential factor in producing both fast and slow wind is the amount by which the field lines spread at increasing heights.

They begin with a model of coronal holes and streamers at solar minimum. Then they inject a spectrum of sound waves and Alfvén waves at the chromosphere. The sound waves shock in the low corona and heat it. Alfvén waves propagate outward and are partially reflected in the corona. Farther out they dissipate in a turbulent cascade. The waves heat the plasma, and the gradient of wave pressure assists in driving the wind.

Their results vary with the shape of the background magnetic field they adopt. In a polar hole, they predict a wind speed of about 800 km/s near Earth.

A maximum temperature of 1.5 million kelvin is reached at 1 solar radius, in good agreement with observations by UVCS. For a model streamer belt, with rapidly diverging field lines outside the streamer, they predict wind speeds, temperatures, and densities that also match results from Ulysses.

What lies in the future of solar wind models? I imagine a merging of the collisionless description (applied, e.g., by Ofman and by Tu and Marsch) with the unified models of Cranmer and friends. Clearly the enormous transverse temperatures and the differences of ion temperatures along the field will have to be addressed with a collisionless theory. But progress to date has been most refreshing after decades of intense intellectual struggle.

A final word about the ultimate fate of the wind in the heliosphere. Observations by space probes Voyagers 1 and 2 showed that the solar wind coasts past all the planets, gradually thinning out in density and magnetic field strength. At some distance, estimated at less than 100 astronomical units (AU) the wind was expected to slow from supersonic to subsonic speeds as it crosses a terminal shock. On December 16, 2004, the space probe Voyager 1 did indeed cross the termination shock at a distance of 93 AU. In December 2006, the craft entered the heliosheath, a region just beyond the terminal shock. Its companion, Voyager 2, crossed the shock at a distance of 84 AU on December 30, 2007. But the true boundary between the wind and the interstellar medium (the bow shock) is thought to lie much farther out, perhaps at 200 AU. If Voyager survives long enough, it may send us a signal from this farthest outpost of our little bubble in space.

THE EARTH'S MAGNETOSPHERE
AND SPACE WEATHER

AT 9:07 IN THE EVENING OF OCTOBER 30, 2003, the entire electric
power grid of southern Sweden shut down. Malmö, a city of 250,000, went
dark in the blink of an eye, as though a giant hand had pulled the plug. Traffic
signals all over the city snapped off, and hospitals went on emergency power.
The city waited several hours for electricity to be restored.

The cause of this blackout was neither a terrorist attack nor an operator's
error. The culprit was the Sun, which had ejected a billion tons of hot magnetized plasma eighteen hours earlier, aimed directly at the Earth. When the
cloud smashed into the Earth's magnetic field, huge electric currents were
induced in the ground. A critical transformer in the grid overloaded, currents
surged elsewhere, and the disaster cascaded throughout the system. The pro-
tective circuitry was too slow to cope. The whole event lasted only a few seconds, but it caused millions of dollars in damaged equipment. The city's utility
has since invested more millions to upgrade its system.

This blackout was minor compared to the March 13, 1989, event, when the
whole electrical grid of the province of Quebec shut down. Six million people
waited nine hours for power to return. Hydro Quebec, the affected utility, has
spent over a billion dollars to protect its grid against future assaults of the
Sun.

Power grids are not the only facilities vulnerable to such storms. They can
damage oil pipelines; disrupt communications with satellites; disorient such
navigational aids as loran and GPS; and interfere with cell phones. To alert

the owners of such delicate equipment, the U.S. government, among others, has established a satellite warning system. The National Oceanic and Atmospheric Administration (NOAA), which operates the National Weather Service and the Hurricane Warning Center, now issues alerts on "space weather" as well. European nations have organized a similar agency, the Space Weather European Network. Private companies have also sprung up to alert special customers.

Fortunately, geomagnetic storms as severe as the 1989 event occur only once in three to five years. More moderate storms occur about ten times a year, on average. But because big storms cause spectacular auroral displays, they have attracted the curiosity of scientists for over a century. For many years, the role of the Sun in causing such storms was controversial, despite mounting empirical evidence. Eventually, the case for the Sun became overwhelming, and the discovery of the solar wind in the 1960s supplied the missing link.

Beginning in the late 1950s, geophysicists have used satellites to probe the extended magnetic field of the Earth and to tease out the intricate processes involved in magnetic storms. As a result of these studies, we have learned just how far the Earth's field departs from a simple dipole and how dynamic the field really is. It's been a revelation.

In this chapter we'll review how scientists gradually learned how the Earth's field is shaped and how it behaves.

THE FIRST HINTS

The story begins with Alexander von Humboldt. He was one of those extraordinary naturalists of the nineteenth century who was interested in everything and who was able to contribute to many different disciplines. Trained as a geologist, he rose in the Ministry of Mines of the Prussian State to the rank of subminister. Upon the death of his mother in 1796, he inherited her fortune and quit the ministry. With his newfound financial freedom, he decided to undertake a private scientific expedition to the Spanish colonies in Central and South America. It was the first of its kind and would last five years. He wanted "to develop an understanding of nature as a whole, (to find) proof of the working together of all the forces of nature."

He was a passionate and indiscriminate observer. He measured everything: the local gravity, the rainfall, the strength of the magnetic field. He collected over 8,000 specimens of plants and animals. He mapped the headwaters of the Orinoco River and the temperature of the ocean currents. (The cold current on the west coast of South America was later named for him.)

On his way home to Berlin via North America, he visited President Thomas Jefferson and lectured at the American Philosophical Society in 1806. Late in life, he wrote a thirty-three-volume treatise *(Kosmos)* summarizing all he had learned of botany, geology, geography, astronomy, and climatology.

In December 1805 he observed wild fluctuations of his compass needle during a brilliant aurora (note 6.1). He gave the name "magnetic storm" to this event, and it marked the beginning of a lifelong interest in terrestrial magnetism. After he was appointed as Prussia's ambassador to France, he had little time for research, but he did build an iron-free magnetic observatory in his embassy's garden in Paris and made regular measurements.

Reports from friends suggested that magnetic storms might be observed over a wide area. To test the idea, he arranged to have observations made in a mine in Freiberg, Saxonia, while he observed in Paris. The results were encouraging but raised more questions. Were the storms a global phenomenon? Were they associated with the position of the Sun, or did they arise from terrestrial effects?

To learn more he would need observations from a widespread grid of magnetic observatories. So when he returned from another expedition, to Siberia in 1829, he convinced the Russian czar to establish a network of magnetic stations across the Russian Empire. Soon he was receiving regular observations of magnetic fluctuations. As we learned in chapter 2, he tried to interest Carl Friedrich Gauss in analyzing them.

Gauss was heavily involved in his mathematical research at the University of Göttingen, however, and picked up Humboldt's suggestion only after the arrival of the young physicist Wilhelm Weber in 1831. For six years these two men collaborated in a series of experiments on electricity and magnetism. Gauss was the theorist of the pair, and Weber the experimentalist.

With Humboldt's encouragement, Gauss and Weber organized a European association of magnetic stations, the Magnetic Union, which published observations of magnetic fluctuations on a regular basis beginning in 1836. Six years of data suggested to Gauss that magnetic storms could indeed be global events, but a network limited to Europe was insufficient to prove that. Humboldt now used his fame and influence to advance the cause. He wrote to the president of the Royal Society of London, a friend he had made at college, and proposed that a chain of magnetic stations be established across the British Empire. As we learned in chapter 2, the British Admiralty adopted his proposal and appointed Sir Edward Sabine as director.

With a decade of observations in hand, Sabine recognized that the frequency of storms tracks the sunspot cycle of eleven years, a definite clue to their origin. But how could the Sun possibly affect the Earth across the 140 million kilometers of empty space that separates them? Most experts scoffed at the idea.

A singular event in 1859 should have convinced them. On September 1 of that year, Richard Carrington was observing the Sun from his London home. He was the son of a rich British brewer. When his father died, Carrington inherited a fortune and was then able to indulge his passion for solar research. He built an observatory, and every clear day he would plot the positions and sizes of the sunspots he could see. In time his accumulated data would lead him to several important discoveries (note 6.2), but none was more important than the one this day.

He had just completed his daily records when he noticed that two bright patches had suddenly appeared near an unusually large group of sunspots. The patches had to be brighter than the surface of the Sun to be visible to his naked eye. They brightened and spread and faded, all within a few minutes. He was astounded. Neither he nor anyone else had seen anything like it. He was, in fact, the first to witness a solar flare in white light. Fortunately, an amateur astronomer was able to confirm his observation.

The real fireworks began eighteen hours later. Telegraph wires all over the United States and Europe shorted out, starting numerous fires. And an enormous auroral display was seen all over Europe, as far south as Rome. Later research revealed that it was also seen in Cuba and Hawaii. Normally auroras are confined to the polar latitudes.

Carrington consulted the magnetic records at the Kew Observatory and learned that the field had decreased sharply at the exact time of the flare and recovered slowly over the following hour. When the aurora began, eighteen hours later, the magnetic record went off-scale. Here was proof that a specific event on the Sun was associated with a large effect on the Earth. Carrington was a cautious man, though, and refused to draw any final conclusions. "One sparrow doesn't make a summer," he said.

Indeed, a possible connection between Sun and Earth was largely discounted even as late as the 1870s. Sir George Airy, the Astronomer Royal, proposed instead that transient electrical currents in the Earth generate countercurrents in the atmosphere, which cause the magnetic storms. Balfour Stewart, director of Kew Observatory, also favored currents but placed them directly in the upper atmosphere.

At least one reputable scientist argued in favor of a solar influence. Henri Becquerel, the discoverer of radioactivity, proposed in 1878 that the Sun emits particles that are responsible for magnetic storms. (George FitzGerald, the Irish mathematician who anticipated some of the predictions of Einstein's special theory of relativity, also suggested particles in 1892.) But at this stage, it was anyone's guess.

Evidence in favor of a solar connection continued to mount. In 1871, Karl Hornstein, director of the observatory at Prague, reported a 26-day periodicity in the frequency of magnetic storms. This period was close enough to the Sun's equatorial rotation period (27.3 days) to suggest a source of some kind on the Sun.

Lord Kelvin, a towering figure in nineteenth-century science, was unconvinced. Kelvin had done fundamental research in the theory of heat and had invented a temperature scale, which was later named for him. He had also proposed with Hermann von Helmholtz that the Sun's source of energy is the heat it acquires by gravitational contraction and that its age could be no more than 10 or 20 million years. This proposal was made, of course, long before atomic energy was discovered.

Kelvin capped a long career by election in 1892 as president of the Royal Society. In his presidential address, he argued forcibly that the Sun couldn't possibly emit enough magnetic energy to disrupt the Earth's field: "During eight hours of a not very severe magnetic storm, as much work must be done by the Sun in sending magnetic waves out in all directions through space as he actually does in four months of his regular heat and light. This result is absolutely conclusive against the supposition that terrestrial magnetic storms are due to magnetic action of the Sun; or to any kind of dynamical action taking place within the Sun, or in connection with hurricanes in his atmosphere, or anywhere near the Sun outside."

As if that were not a sufficiently strong statement, he added, "The supposed connection between magnetic storms and sunspots is unreal, and the seeming agreement between the periods has been a mere coincidence" (*Publication of the Astronomical Society of the Pacific* 5, 45, 1893).

More convincing proof was about to appear. E. Walter Maunder, at the Royal Greenwich Observatory, discovered in 1904 that the greatest magnetic storms occur when a group of large sunspots lie near the center of the solar disk. More importantly, the storms recur as the same spots rotate back to the Sun's center. Even after the spots disappear, the storms continued to recur, as though the area the spots had occupied was still active in some way. Maunder suggested that the area around a sunspot group (an M region) emits a stream of some sort, which rotates with the Sun (like a lawn sprinkler) and periodically sprays the Earth. From the duration of a storm, he estimated a width of about 20 degrees for the stream.

To many scientists, Maunder's results were sufficient to overcome Kelvin's objections. Some special areas on the Sun seem to be the sources of streams that intersect the Earth periodically and produce magnetic storms, but the origin and composition of the streams would not be known for another seventy years.

All this debate was going on with no information on the magnetic field of the Sun. Astronomers had seen feathery plumes in the solar corona during total eclipses but couldn't agree on whether they might be the traces of a solar magnetic field. Indeed, some thought the corona itself was only an effect of the terrestrial atmosphere. Then in 1908, George Ellery Hale, director of the Mount Wilson Observatory, discovered that sunspots contain magnetic fields thousands of times stronger than Earth's (see chapter 3). Hale agreed with Kelvin, however, that the Sun's magnetism, by itself, isn't strong enough to cause magnetic storms on Earth. He favored some influence from solar flares, without being too specific.

Streams of particles, or "corpuscles," were now considered the most likely agent for producing magnetic storms. What could these particles be? Kristian Birkeland, a Norwegian physicist, approached the question as an experimentalist. In the early 1900s he had organized a number of expeditions to northern Norway, where he established a network of four magnetic stations in the auroral zone. From these observations he determined that the auroras arise from electrical currents that stream along the magnetic lines of force. He speculated that these field-aligned electric currents (later called electrojets) are part of a circuit that links the auroral zone to lower latitudes, but the identity of the particles that carried the electrical currents remained open.

Birkeland thought that electrons might be the particles in question. So in the early 1900s, he set up a series of laboratory experiments at the University of Oslo to test this hypothesis. He built a magnetized sphere (a terrella) and coated it with a fluorescent paint. Then he fired electrons at it within an evacuated vessel. In his experiments of 1903 and 1905, he was delighted to see the poles of the terrella light up, just where the electrons were funneled inward by the field lines.

The similarity to auroras was striking, and he thought he had identified the unknown particles. But there were several problems with his explanation. First, auroras don't occur directly over a magnetic pole but around an oval several thousand kilometers in diameter that is centered on that pole (note 6.3). Moreover, electrons couldn't continue to leave the Sun indefinitely; a large positive charge would build up to restrain them. A stream would have to have electrons and protons in equal numbers.

For these reasons, Birkeland's conclusions on the origin of auroras were mostly ignored or disputed. Only after satellites began to probe the magnetosphere, fifty years after his death, were his basic ideas on auroral currents confirmed.

The ridicule he endured affected his personality. Even as a young man he was a solitary and reserved character, with an almost obsessive intensity in his research. When his work on auroras was criticized, he became even more eccentric. For example, he conducted his laboratory experiments dressed in a red fez and pointed Egyptian slippers, a costume hardly likely to escape notice. But he was also a practical man. To fund his experiments, he invented an electromagnetic cannon and tried to sell it to the military. Unfortunately, his prototype exploded in its first test, forcing him to abandon the project. But this venture led to a profitable method of extracting nitrogen from the air to produce fertilizers. You can't keep a good man down.

THE FIRST THEORIES

Thirty years passed without significant progress. Then in 1931 Sydney Chapman and his student Vincent Ferraro proposed the first theory for the interaction of a neutral particle stream and the Earth's dipole field.

Chapman was one of the most productive geophysicists of the twentieth century. He made important contributions to ionospheric and magnetospheric physics; the kinetic theory of gases; aeronomy; and meteorology. He graduated from Cambridge University in 1910 as a theoretical physicist. Shortly before World War I, in a paper of fundamental importance, he showed how to calculate the diffusion of heat or particles in a gas from first principles. Later he completed important work on plasmas and gas dynamics. He also advanced a theory, which turned out to be wrong, of geomagnetic variations based on the tides.

In 1931, when Chapman and Ferraro proposed their theory of magnetic storms, practically nothing was known about solar streams, so they had to make some assumptions. They assumed that the stream's diameter was much larger than the Earth and had a flat front face. As the stream hit the Earth's dipole field, electrical currents would be induced in the stream, which would generate a secondary field. The stream's field would compress the Earth's field on the side facing the Sun, like an advancing snowplow. A small, sharp increase in the field at the Earth's surface would therefore occur, as indeed had been observed at the onset of magnetic storms.

The Earth's field would resist being compressed, which would slow the stream, most effectively on the sunlit hemisphere. As a result the stream would slow down near the equator but elsewhere would continue to flow unimpeded around the Earth. In effect the stream would *engulf* the Earth's dipole field, closing in behind it to form a "cavity" in the stream. (Imagine a tethered balloon in a windstorm.)

Opposite sides of the cavity would become polarized, with ions on one side and electrons on the other. An electric current would then flow in a ring that circles the equator at a distance of a few Earth radii. The magnetic field of this current would oppose the Earth's field, and that would account for the observed 2 or 3 percent decrease in the field strength at the surface. As the stream passed the Earth during a day or two, the ring current would fade and the field at the Earth's surface would slowly increase to its normal value. The storm would have passed.

This early theory of magnetic storms was an important step toward explaining the sequence of events that was actually observed. However, Chapman and Ferraro didn't consider that the Sun might emit a *continuous* stream of plasma; instead they thought of discrete clouds. Nor did they examine what might occur if a solar stream contained an embedded magnetic field. Finally, to make life simple, they assumed the stream was a perfect conductor of electricity; it had zero resistivity. That meant that field lines were frozen into the plasma.

Another twenty years passed without much progress. Then in 1958, two milestones were reached: James Van Allen discovered "radiation belts" near the Earth (note 6.4) that resembled the predicted ring currents of Chapman and Ferraro; and Eugene Parker published his predictions of a steady solar wind and a spiral interplanetary magnetic field (IMF; see chapter 5).

James Dungey, a theoretical physicist in Britain, and a former student of Fred Hoyle, immediately recognized that such a magnetized solar wind could interact with the geomagnetic field in a different way than Chapman and Ferraro had proposed. How it interacted would depend on the direction of the wind's magnetic field. As conditions on the Sun change, so too would the direction of the interplanetary field. Half the time the field lines might arrive at Earth tilted slightly toward the north, half the time tilted toward the south. Dungey realized that a south-pointing component could *reconnect* with the Earth's north-pointing field. That might have enormous consequences.

Ronald Giovanelli had originated the concept of magnetic reconnection in 1947, in an attempt to explain solar flares (see chapter 3). In 1958, Paul Sweet, at the University of London Observatory; Eugene Parker, at the University of Utah; and James Dungey, at Cambridge University, all published independently on the subject. They analyzed the behavior of two opposite-polarity fields that are embedded in plasma when the fields are squeezed together. They predicted that a plasma layer only a few meters in thickness develops between them, in which a powerful electric current flows. If the layer was not perfectly conducting, field lines would no longer be frozen-in but could diffuse and reconnect with lines of opposite polarity. Somewhere within the layer a "neutral point" would exist, where opposite fields would cancel to zero. Reconnection of fields changes their overall shapes and also can release stored magnetic energy. In solar flares, this energy is converted to heat and particle kinetic energy.

Dungey adapted the idea of reconnection to explain geomagnetic storms and auroras. In his original model Dungey pictured a steady solar wind at a time when the embedded interplanetary field pointed to the south (fig. 6.1 top). We can follow the evolution of a typical interplanetary magnetic line (labeled 1 in the figure) in the northern hemisphere. Imagine the solar wind blowing in from the left of the figure.

The wind pushes the line into contact with the Earth's north-pointing field, as at position 2. A neutral point (N) would develop near the equator, and the line would connect with the terrestrial line 2'. The newly connected

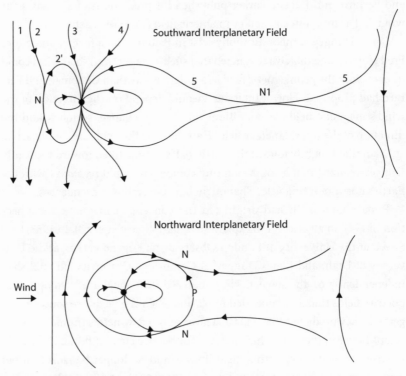

Fig. 6.1. Dungey's open field model of magnetic storms. He proposed reconnection between the interplanetary and terrestrial fields at two neutral points, labeled N and N1.

line would be severely kinked near the neutral point, and its internal tension would force it to straighten up. Therefore, the kink would unbend rapidly toward the north magnetic pole, carrying plasma with it.

As the solar wind continues downstream, the reconnected line (3) is swept to position 4 but keeps one end firmly anchored near the north magnetic pole. (Dungey showed, in fact, that such a line is rooted in the auroral oval.) Continuing on, the "open" end of the line wraps around the Earth's dipole field, as in position 5. As more interplanetary field lines wrap around the dipole, the lateral magnetic pressure builds up. As a result, the line we are following is gradually forced inward. A long pointed tail begins to develop in the magnetosphere.

The sequence of events is symmetrical south of the equator, so our chosen line would meet its opposite-polarity mate at a second neutral point labeled N1. When these two lines reconnect, a teardrop-shaped loop would form under high tension. This stressed loop would relieve its tension by snapping back toward the Earth like a stretched rubber band, carrying plasma. It would now join the original dipole field, which is rotating in pace with the solid Earth.

As part of the reconnection event in the tail, another kinked line forms downstream of the neutral point N1. It too would straighten out and detach from the Earth's tail to float off downstream with the solar wind.

Dungey introduced several radically new concepts in this model. He employed reconnection for the first time, not only at the sunlit side of the magnetosphere, but a second neutral point in the tail. He predicted that solar plasma and a magnetic field would circulate (or "convect") from the neutral point on the day side to the second neutral point in the night side and back again. Each collapse of the stressed teardrop loop would accelerate plasma, which could funnel energetic particles into the auroral zones. In his scenario, therefore, the aurora arises from tail particles, not directly from the solar wind. In this whole scenario of reconnection, merging, and plasma circulation, the wind supplies both energy and mass to the magnetosphere.

This version of the model makes sense if the interplanetary field happens to point south. But suppose it points north? Then what happens? Dungey extended his model in 1963 to cover this important situation (fig. 6.1 bottom). A north-pointing field line couldn't reconnect at the day side of the magnetosphere (the "nose"), so its open ends would be swept downstream to fold over the Earth's dipole field. On the night side of the Earth, the interplanetary field line would then contact dipole field lines. In both northern and southern hemispheres, these dipole field lines have the opposite polarity to the open ends. It is there on the night side flanks of the dipole (labeled N) that reconnection could occur. Dungey proposed that such events account for the weak substorms that are observed. We'll return to these later.

Strong evidence in support of Dungey's "open field" model appeared in 1966, when it was realized that magnetic activity and auroras are far more frequent when the interplanetary field tilts south. Starting in 1958, with James Van Allen's experiment aboard Explorer 1 (note 6.4), a series of satellite experiments has filled in our picture of the Earth's field and has confirmed most of Dungey's vision. In figure 6.2 we see the current model. It is overloaded with new names, for which I apologize. The most important name to remember is "magnetosphere," first applied to the Earth's extended field in 1959 by Thomas Gold, a Cornell University astrophysicist.

The Earth's magnetosphere obstructs the free flow of solar wind, like a rock in a river. The wind, blowing at over 300 km/s, "senses" the presence of the magnetosphere by forming a steady shock front, analogous to the wave at the bow of a cruising ship or at the nose of a supersonic plane. The bow shock was discovered at 10 to 15 Earth radii upstream by the satellite IMP 1

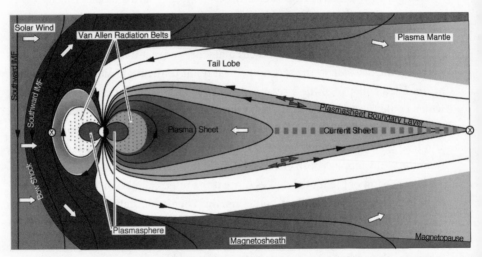

Fig. 6.2. This sketch shows the many different parts of the Earth's magnetosphere, as revealed by a series of satellite observations.

in 1966. Just upstream from the bow shock, the interplanetary field strength ranges between one and a few hundredths of a microgauss, or about a thousand times weaker than Earth's field at the magnetic pole.

The wind slows and heats up as it passes through the bow shock. It continues to flow around the closed field lines of the magnetosphere in a region named the magnetosheath. Between the magnetosphere and magnetosheath lies a thin boundary layer called the magnetopause, which was discovered by Explorer 12 in 1962.

The teardrop tail of the magnetosphere, which was detected by IMP 1 in 1964, fluctuates in length. Sometimes it stretches a thousand Earth radii downstream. But in 1983, for instance, ISEE 3 (International Sun-Earth Explorer) discovered plasma escaping to space beyond the tip of the tail, at only 70 radii. For comparison, the Moon's average distance is 60 radii.

North of the equatorial plane, the tail's field points toward Earth; below the plane it points away. Separating these two regions is a plasma sheet, which has a temperature of about 5 million kelvin and density of 1 proton/cm^3.

Within 2 Earth radii (geocentric) lies the plasmasphere, a volume that contains plasma denser than 100/cm^3 and cooler than about 10,000 K. This cold plasma rotates with the Earth. When it reaches the reconnection zone on the day side, it escapes from the dipole field, merges with the solar plasma, and is replaced eventually by ions from the ionosphere.

The dipole field between 2 and 10 radii has a doughnut shape. It contains two "radiation" belts filled with trapped particles. Van Allen discovered the inner belt between 1 and 3 radii in 1958 (see note 6.4). Here are protons with energy higher than 10 million electron volts that are produced by impacts of galactic cosmic rays on the atmosphere. These hot protons reach altitudes as low as 250 km in the South Atlantic Anomaly, a region of strong field near Brazil, where they can damage satellites in low orbits.

The outer radiation belt, between 3 and 9 radii, contains relativistic electrons with energy as high as 10 million electron volts. They are accelerated in geomagnetic storms, to which we now return.

MAGNETIC WEATHER

Storms come in several varieties, and our first task is to sort them out. At one end of the scale are the superstorms like the 1989 event that blacked out

Quebec. These great magnetic disturbances are observed globally and are accompanied by auroras that extend to low latitudes. They may last for several days. The first sign is a short, sharp jump in the field strength at the surface, followed by a slow decrease of perhaps 0.2 percent and a gradual recovery over hours or days. Such storms are rare, but during the declining phase of the eleven-year sunspot cycle, more moderate storms occur once or twice a month. We now know that they are caused by coronal mass ejections or by interplanetary shocks ejected from solar flares. Typically a mass ejection arrives at Earth after a day or two.

Recurrent storms are entirely another breed. They are caused by high-speed wind streams that originate in coronal holes or by the shocks they create when they overtake slower streams. The streams rotate with the Sun in about twenty-seven days and may persist for months. With each rotation a stream with a south-pointing internal field can sweep past the Earth and cause a storm.

Around 1961, Sydney Chapman and Syun-Ichi Akasofu described a third type of storm, an intense but brief magnetic disturbance associated with the brightening of auroras. They named them substorms. In a typical substorm the field strength near the magnetic pole dips by several percent (ten times larger than in a major storm) and recovers in about an hour. Another substorm may occur within a few hours.

Chapman thought that all major storms were made up of clusters of substorms, but that view proved to be too simple. Akasofu proved that isolated substorms occur, at least in the auroral zones, whenever the interplanetary field turns south. At first, substorms seemed simpler to analyze than major storms, so geophysicists focused on them. As more satellite data accumulated, the picture became very complicated indeed. At least six different models of the evolution of substorms have been proposed in the last decade, no consensus has been achieved, and several key questions remain unanswered. I offer here a simplified description of a typical substorm based on the 1996 Near Earth Neutral Line model of Daniel N. Baker and associates.

An isolated substorm moves through three distinct phases: growth, expansion, and recovery (fig. 6.3). The growth phase (A) begins when the interplanetary field turns south and lasts about an hour. Reconnection starts at the nose of the magnetosphere and sets off a transfer of magnetic flux from the day side dipole field to the tail lobes. As magnetic pressure builds up at the magnetopause, the lobes are compressed and a reconnection site appears, as in Dungey's scenario. The tail continues to grow in length and in field strength for some tens of minutes. Magnetic energy is stored during this time and is released in the following expansion phase.

Suddenly, and for controversial reasons, a new reconnection site develops near the Earth, at a distance of about 30 radii (fig. 6.3B). The closed loops that form in the reconnection snap back toward the Earth like stretched rubber bands, heating and compressing their trapped plasma (fig. 6.3C). Electrons flood down the polar field lines, bombard the atmosphere, and create a brilliant aurora at heights above 100 km. These electrons continue to flow as strong electric currents in the ionosphere, generating the magnetic disturbance one observes at ground level.

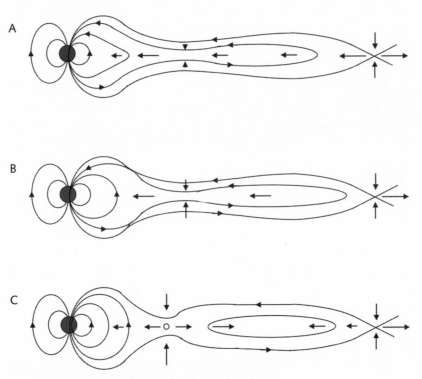

Fig. 6.3. The evolution of a substorm takes place in three phases.

Images of the northern auroral zone, obtained by the IMAGE satellite, show that the aurora begins near the midnight meridian and expands both east and west along the auroral oval as well as toward the magnetic pole.

In the middle part of the tail, a bubble (or plasmoid) of magnetized plasma pinches off and roars downstream at speeds as great as 800 km/s (fig. 6.3C). After a few tens of minutes the disrupted tail begins to recover to its initial state. The substorm is fading away. Full recoveries make take an hour or more.

MAJOR STORMS

A major geomagnetic storm, like the 1989 event, usually requires a trigger, like an interplanetary shock or coronal mass ejection. But as with a substorm, a south-pointing interplanetary field is also a necessary condition.

The key difference between substorms and major storms is the role of the ring current. Recall that Chapman and Ferraro postulated that during a geomagnetic storm, a transient current would be induced in an equatorial ring around the Earth. The magnetic field of the ring current would oppose the dipole field and cause its observed decrease.

We now know that a *permanent* ring current exists in the outer radiation belt of the dipole, between 2 and 7 radii. This region normally contains a small number of trapped energetic charged particles (10 to 200 keV). The particles spiral along closed field lines and bounce back and forth between the magnetic poles. A particle can jump from one closed field line to another. Ions jump westward, electrons jump eastward, and together they constitute the permanent westward ring current. The current generates a weak magnetic field that opposes and weakens the Earth's dipole field.

When a coronal mass ejection arrives at Earth, it compresses the nose of the magnetosphere suddenly, which induces large currents in the ionosphere and ground. It also injects a large amount of solar plasma into the plasma sheet, via the magnetotail, by a rapid series of intense substorms. When the bloated plasma sheet expands Earthward, its electrons and protons jump into the closed field lines of the dipole within 4 radii, which greatly enhances the ring current. As a result, the ring's magnetic field increases, and being opposite in polarity, it weakens the Earth's existing field. It's this change in field strength, a few tenths of a percent over six hours, that causes all the damage to sensitive equipment in space and on the ground.

We've just finished describing the large-scale features of the magnetosphere and its interactions with the solar wind. Geophysicists pieced together the picture we've presented, in rather coarse detail, during the decades of the 1960s and 1970s. Certainly by 1985, the major discoveries in the field had been made, and progress in the next decade slowed visibly. In the early 1990s, prominent workers in the field, such as Christopher Russell at UCLA and David Stern at NASA's Goddard Space Flight Center, focused attention on the next round of problems to be solved and suggested strategies for solving them.

Russell and Stern both recognized that observations from one or two satellites would no longer be adequate to crack some of the more difficult problems, such as the genesis of substorms or the detailed physics of reconnection. Russell had had personal experience using two satellites. During the International Magnetospheric Study of the late 1970s, his magnetometers aboard the satellites ISEE 1 and ISEE 2 had provided the first glimpses of "impulsive, patchy reconnection" at the day side of the magnetosphere (C. T. Russell and R. C. Elphic, *Geophysical Research Letters* 6, 33, 1979). But to untangle the complicated processes during reconnection, more fine-grained observations would be needed.

So Russell called for *fleets* of satellites, numbering four or more, that could observe the same event simultaneously from different vantage points. At the same time, he stressed the importance of developing mathematical models to describe observations already in hand and mining old data to find simple examples to work on. Both of these approaches have been pursued and have proved their worth.

SATELLITES OF THE 1990S

Satellites are notoriously expensive to build and operate, and yet they are the primary tools for probing the magnetosphere. Their high cost drove NASA, the European Space Agency (ESA), and the Japanese Institute for Space and Aeronautical Science (ISAS) to collaborate in a program in the late 1980s called the International Solar-Terrestrial Physics (ISTP) initiative. Their prime goal is to develop a better understanding of the impact of solar plasma and fields on the Earth's environment.

Beginning in 1992, the agencies launched three satellites, Geotail, Wind, and Polar. The spacecraft operate independently, but the observations they make are often complementary. Geotail was built by ISAS and launched by NASA in July 1992 into an elliptical orbit that ranged between 8 and 200 Earth radii. ISAS also designed and built five of the seven instruments that sample the plasma, electric and magnetic fields, and particle composition. During its decade-long flight, Geotail revealed complex processes of particle acceleration by electrostatic and plasma waves in the distant tail. The details of these processes are too technical to go into here. Suffice to say that Geotail's observations have stimulated wide-ranging theoretical studies and simulations that are relevant elsewhere in the universe.

Wind was launched in 1994 to sample the solar wind upstream from the Earth. It carried nine clusters of instruments to observe all the properties of the wind as well as gamma ray bursts from the Sun. Its operations closed down in 2003.

Polar was launched in 1996 into an orbit highly inclined to the equator, to enable it to view the poles nearly continuously. Its instruments were designed to observe the interaction of the wind and the ionosphere, the production of auroras, and the flow of plasma into the cusp. Polar continued to observe until 2007.

The ISTP program can claim a number of important contributions. I picked three. Perhaps the most far ranging is a shift in paradigm: the geotail is not the most important element in the activity of the magnetosphere. Rather it appears that the critical processes occur much closer to Earth, probably in the region of the ring current.

Second, and contrary to intuition, some shock fronts in the solar wind can arrive at Earth *before* they are observed at the L1 Lagrangian point, some 1.5 million kilometers from Earth. That can thoroughly confuse modelers who are trying to understand the subsequent events.

Third, collisionless reconnection (note 6.5) is now recognized as the most important agency for transferring energy from the wind to the magnetosphere, and it occurs in more locations than previously thought. As we shall see, reconnection occurs throughout the universe in many different contexts, so that understanding its details is essential. The magnetosphere is an excellent laboratory for this purpose, because satellites can enter the reconnection region, in contrast to the Sun, where reconnection can only be observed indirectly.

RECONNECTION IS HOT!

ISTP satellites detected the telltale markers of reconnection on several favorable occasions. On April 1, 1999, for example, the Wind satellite chanced to pass through the geotail for twenty minutes while reconnection was occurring. An instrument built by a team at UC Berkeley's Space Sciences Laboratory detected a burst of electrons accelerated to 80 percent of the velocity of light, which corresponds to energy of 300,000 electron volts.

Marit Øieroset, a member of the team said, "With Wind, we have been very lucky. It went through the magic region in the magnetosphere where field lines actually reconnect and saw the process in action." Robert Lin, another team member, agreed: "This observation was the first clear-cut, unambiguous evidence that a region of magnetic reconnection is the source of high-energy electrons" (*NASA Science Daily,* July 26, 2001).

The Berkeley group concluded that the reconnection region was about half the size of the Moon. In addition their data confirmed a prediction that reconnection involves the *collisionless* diffusion of ions and electrons away from lines of force. The ions leave the field lines before the electrons, which remain to guide the lines to the point where they can reconnect. Collisionless reconnection can occur much faster than the more familiar collisional reconnection that occurs in the Sun (M. Øieroset et al., *Physical Review Letters* 89, no. 19, 2002).

Exactly two years later, on April 1, 2001, the Geotail and Wind satellites happened to observe the reconnection region at the nose of the magnetosphere, where the IMF first contacts the Earth's field. Wind was located at the dawn side of the nose, while Geotail swept through the magnetopause for several orbits. In its multiple crossings, Geotail observed a reversal of the field's polarity, which the Berkeley team interpreted as the signature of an X-type reconnection site (note 6.6).

CLUSTER

Since the pioneering work of Dungey, geophysicists were accustomed to thinking about reconnection as a two-dimensional process between north- and south-pointing field lines, as in figure 6.1. When the IMF direction is arbitrary, however, reconnection becomes a three-dimensional process that is only beginning to be understood. Everyone realized that the magnetosphere could provide valuable insights into the process but only if the observations were finer-grained. That would require observations with more than two satellites.

Way back in 1982, the ESA, with headquarters in Paris, began to plan a fleet of four spacecraft, called Cluster, that would fly in formation to make coordinated observations. Their mission would be to explore the cusp and magnetotail regions in three dimensions, with time and space resolution beyond anything achieved before.

The four satellites were built and instrumented in Europe and ready for launch by 1996. In June of that year, Cluster was placed on an Ariane rocket at Kourou, French Guiana. The rocket fired and rose majestically, but within seconds after liftoff, it broke up and had to be destroyed. Cluster was a total loss.

ESA decided the mission was too important to give up, so four new satellites were built from spare parts, and in 2000 they were launched in pairs with two Soyuz rockets from the Baikonur Cosmodrome in Kazakhstan. The satellites cycle over the poles every fifty-seven minutes in an elliptical orbit that ranges from 19,000 km (3 R, or 3 Earth radii) to 119,000 km (19 R). Each satellite carries an identical set of eleven instruments to measure charged particles as well as electric and magnetic fields.

Cluster has been performing beautifully since 2000, and its mission has been extended to December 2009. It has a long list of science accomplishments. I've chosen a few important discoveries, all relating to reconnection.

On September 15, 2001, a large team of Chinese, European, and American researchers observed a magnetic "null point" in the geotail for the first time. A null point is a small region where ions and electrons diffuse away from field lines and allow the lines to break and reform. In two dimensions, the null point lies at the center of an X-shaped arrangement of field lines (see fig.

measure plasma properties, was orbiting about the L1 Lagrangian point, 236 R upstream from Earth.

Over a period of several hours, an enormous reconnection structure in the solar wind swept over all six satellites at a speed of 340 km/s. At ACE, the polarity of the IMF reversed abruptly, and instruments onboard sensed a powerful jet of plasma directed toward the equatorial plane. Using observations from all six satellites, Tai Phan at UC Berkeley and John Gosling at the University of Colorado were able to reconstruct the X-type shape of the event.

In their model, two flat sheets of oppositely directed magnetic field intersected hundreds of Earth radii north of the equatorial plane. When viewed in cross-section, the sheets would look like the arms of the letter X and they would cross at a so-called X-line. The length of this line, the actual site of reconnection, was estimated at a staggering 390 Earth radii, or 2.5 million km.

The investigators observed twenty-seven such events with somewhat less detail. They concluded that reconnection in the wind is fundamentally a large-scale process. The patchy, impulsive reconnection that is observed in relatively small regions of the magnetosphere is probably caused by fluctuating conditions at the magnetopause and is not fundamental to reconnection itself.

Perhaps the most important result of this research was the realization that reconnection can be maintained in the freely flowing wind without being forced by a pressure imbalance.

Reconnection is by no means the only topic being investigated with the Cluster satellites. Here is another example. Researchers have been trying for years to explain why the magnetosphere contains three to five times more plasma when the IMF points north than when it points south. One would think that a south-pointing IMF would be more effective in loading the magnetosphere, because reconnection with the Earth's north-pointing field is then easier.

On November 20, 2001, Hiroshi Hasegawa and colleagues at Dartmouth College obtained observations from Cluster that help to answer this puzzle. Their discovery depended on having two satellites of Cluster on each side of the magnetopause.

The observations reveal the growth of vortices at the magnetopause that inject solar wind plasma into the magnetosphere. The vortices resemble the curling "pipeline" waves that surfers ride along the face of a breaking ocean wave. They are huge—more than three times the diameter of Earth.

Hasegawa explained that they arise from the friction between the fast solar wind and the flanks of the magnetopause. As the wind streams by, ripples and waves build up on the surface of the magnetopause, in the same way that ripples develop on the surface of the ocean in a fresh breeze. The ripples grow in amplitude as they move downstream. When they finally "crash" into vortices near the geotail, they capture solar plasma. Evidently, such vortices develop preferentially during periods of north-pointing IMF, but the reason why is still uncertain.

SPACE WEATHER

Does all this basic research have any relevance to the practical problem of, say, avoiding power blackouts? The researchers certainly hope so and are working hard to incorporate their latest findings into predictive models. A good example of such an effort is the GEM program at UCLA. GEM stands for Global Environment Modeling. Beginning in 1991, a large international group secured funds from the U.S. National Science Foundation to carry out theoretical and analytical research in a series of campaigns. Each campaign would target a cluster of problems associated with a selected region or process in the magnetosphere. The first, from 1991 to 1996, focused on the physics of the cusp and boundary layer. The second campaign, running from 1994 to 2003, tackled the geotail and substorms. The latest work relates to the inner magnetosphere and coupling to the ionosphere.

The ultimate goal of the program is to construct a Global Geospace Circulation Model, capable of predicting the evolution of the magnetosphere. It would be analogous to the general circulation models that atmospheric scientists run on giant computers to forecast weather and trends in climate. If attainable, such a circulation model could be used to forecast major geomagnetic storms in the same way that the Weather Service forecasts the evolution of hurricanes.

Working toward this distant goal, the members of the GEM program have investigated literally dozens of aspects of the physics of the magnetosphere and have published hundreds of scientific articles. The power of the team approach lies in the close interaction among theorists, data analysts, and modelers. Consider the Grand Challenge issued in 1998 by the data analysts to the modelers. Could the existing computer models reproduce the observed reaction of the polar ionosphere to different orientations of the interplanetary magnetic field? Data from four orbiting Defense Department satellites were compared with five different models. None was entirely successful, but one came close. From such exercises, the modelers get clues on how to fine-tune their computer programs.

Some modelers, like Joachim Raeder at UCLA, are building computer programs to simulate the evolution of a complete substorm, a really ambitious goal. In 2003, Raeder summarized the techniques and status of such research (in *Space Plasma Simulation,* ed. J. Buchner, Springer, 2003, 212–46). After one typical test case he concluded, somewhat pessimistically:

> The simulation also shows why data analysis has not yet provided a clear picture (not even a phenomenological one) of the sub-storm process and why intelligent people might come to quite different conclusions by looking at essentially the same data sets. Figures (x and y) show that even a small variation in the geotail's position leads to significantly different observations.
>
> One might be tempted to conclude that studies using observations from a single (or a few) spacecraft are doomed to fail to solve the problem because they can never derive suitable synoptic maps and at best murky statistics. Thus, a convincing solution may only be found with constellations of 10's to 100's of spacecraft, along with modeling and data assimilation.

Such large fleets are unlikely to materialize soon, but the scientists who forecast impending magnetic storms are doing quite well with four or five. Let's see how they do it.

THE STORMS OF 2003

The U.S. Space Environment Center (SEC) in Boulder, Colorado, is the nation's front line of defense against space weather and is the official source of alerts. Deep inside the center, teams of forecasters watch a bank of computer screens twenty-four hours a day. Their job is to evaluate the stream of solar, interplanetary, and magnetospheric observations they receive from ground-based and satellite observatories. They interpret the data with the aid of computer models and publish summaries on the Internet. If necessary they issue daily or hourly warnings of impending threats. In the autumn of 2003, they showed what they could do.

The Sun went wild in late October and early November. It erupted with the most powerful flares ever recorded and caused a geomagnetic storm that rivaled the 1989 event. The intensity of activity was all the more astonishing because sunspot numbers had peaked two years earlier and the Sun had been quieting down ever since.

Forecasters at SEC received the first signs of trouble from several ground-based solar observatories on October 18. They reported that a rapidly growing pair of sunspots in Active Region 484 was just rotating on to the visible solar disk. By the 19th, the spots covered an area ten times the size of the Earth and the region emitted a weak x-ray flare. SEC received the x-ray data within minutes from its Geostationary Operational Environmental Satellite (GOES), which orbits around the L1 Lagrangian point.

By the 20th the active region was large enough to be seen through dark glasses with the naked eye, and by the 21st, it grew to one of the largest ever seen. The Michelson Doppler Imager (MDI) instrument aboard the SOHO satellite revealed powerful magnetic fields linking the central spots, in a "delta" configuration known to produce big flares. The signs were ominous.

On October 22, the LASCO coronagraph aboard SOHO transmitted images of a relatively slow CME to the SEC forecasters. They predicted correctly that the cloud would arrive on the 24th, and would cause only a moderate geomagnetic storm, rated G3 on a scale of 1 to 5.

The best was yet to come. A new active region, labeled 486, rotated on to the solar disk on October 26. It too had two large sunspots within the same penumbra, a delta configuration. The region warmed up with a class X2 flare on the 25th and a CME that arrived two days later. SOHO detected the CME as it left the Sun, and SEC received details of its speed and density as it passed the ACE satellite. This Advanced Composition Explorer has been sampling the solar wind plasma from its orbit around the L1 Lagrangian point since 1997 and has been an invaluable asset to SEC ever since. Despite the warning that SEC published, the Japanese Earth Observing Satellite Midori 2 was caught unaware and crippled by this well-positioned CME.

Region 486 continued to grow and was fifteen times the size of Earth when it reached the center of the solar disk on October 28. It was now the largest spot group of this solar cycle and was ready to show its power. Forecasters at SEC were worried that its delta configuration could produce a powerful flux of high-energy protons if it flared. They warned the astronauts aboard the International Space Station to take cover in the most protected modules.

Later that day the region erupted with a class X17 x-ray flare, the third largest ever seen. As predicted, the flare also produced a dangerous proton burst, which was registered within minutes by ACE and then by GOES. The flux of 10-MeV protons in this event was the second largest of this cycle. These lethal protons bombarded the Japanese Kodama satellite and fried its electronics instantly. Had the operators taken precautions in time, they might have saved the craft.

At the same time as the flare, observers at the SOHO operation center were awestruck as Region 486 ejected a huge CME. The cloud passed ACE after a mere nineteen hours, one of the fastest transits of the present cycle, and slammed into the Earth's magnetosphere. SEC had been transmitting urgent alerts to the world for the past twenty-four hours. Forecasters had taken care to warn the Federal Aviation Administration (FAA), which rerouted planes to avoid latitudes above 57 degrees, where high-frequency communications are often disrupted by storms. Now the storm was upon us.

The biggest geomagnetic storm of the cycle, rated as a G5, began on the 29th and lasted for over a day. Auroras were seen as far south as Spain. As expected, power grids in the northwest United States and in Europe experienced violent voltage swings, and strong induced currents were induced globally when the CME's shock arrived.

The region exploded with a second large flare, rated X10, on October 29 and ejected yet another CME. After a short nineteen-hour flight, the cloud passed ACE on the 30th and caused another G5 geomagnetic storm. This time Malmö, in Sweden, was blacked out for several hours.

The climax to the drama occurred on November 4, when Region 486 emitted the largest x-ray flare ever observed, rated a stunning X28. Fortunately, the region was far over toward the west edge of the Sun and its powerful CME blew off harmlessly away from Earth.

When the fireworks subsided, the experts gathered to assess their success in forecasting during this exciting period. On the whole the system worked as well as intended. Damage to vulnerable systems was minimal. "I think this proves that the warning system we have works," said Paal Brekke, deputy project scientist for SOHO. Joe Kunches, head forecaster at the SEC was also pleased.

John Kapperman, president of the private forecasting company Metatech, pointed out that we were all very lucky that these huge geomagnetic storms didn't cause more damage. The CME that caused the G5 storm of October 28–29, for example, contained a north-pointing magnetic field and therefore was more subdued than it might have been. Moreover, enough time elapsed between the storms of October 28 and October 30 for the magnetosphere to recover somewhat; otherwise the two storms might have merged into a colossal superstorm. Nevertheless, the combined forces of ground and space instruments, coupled with the insight of the forecasters, proved a match to this violent episode.

We can only hope that our skill in dealing with space weather continues to improve as the Sun moves toward its next maximum of activity in 2012.

THE PLANETS

FOUR AND A HALF BILLION YEARS AGO, the Sun and its family of planets were born from a giant cloud of interstellar gases. As scientists reconstruct the event, the slowly spinning cloud collapsed under its own gravity and spun up to form a flat disk. Most of the mass collected at the center, in a warming protostar that would become the Sun. Clumps of frozen gases in the disk drew together into small bodies, planetesimals, which in turn collided and grew larger. Gradually the disk coalesced into nine planets and their many moons.

After about 100 million years, thermonuclear reactions ignited in the Sun and sunlight streamed out to warm the disk. Light gases, such as hydrogen, helium, and methane, heated up sufficiently to escape the planets nearest the Sun. Later, a vigorous solar wind stripped the terrestrial planets (Mercury, Venus, Earth, and Mars) of most of their mass. They were left as small bodies of molten silicates and metals. Farther from the Sun, the planets remained cold enough to retain most of their mass. The giant planets Jupiter, Saturn, Neptune, and Uranus were born. Pluto's origin is still controversial (note 7.1). Beyond the planets a chaotic cloud of icy bodies was left over as a spawning ground for comets.

The original cloud probably possessed a feeble disorganized magnetic field, similar in strength to those detected in interstellar clouds today. As the Sun and planets formed, they each retained some of this field. As they evolved further, they amplified these "seed fields," probably by dynamo action in their interiors. But for reasons that are still controversial, Mercury, Venus, and

Table 7.1. Properties of the planets

Planet	Radius[a]	Mass[a]	Rotation	Axis Tilt (degrees)[b]	Field[c]
Mercury	0.38	0.055	58.7 days	0.1	0.0025
Venus	0.95	0.82	243 days	−2	—
Earth	1.0	1.0	23.9 hrs	23.5	0.32
Mars	0.53	0.11	24.6 hrs	24	—
Jupiter	11.2	318	9.92 hrs	3.1	4.4
Saturn	9.46	95.2	10.7 hrs	29	0.20
Uranus	4.1	14.5	−17.3 hrs	−82	0.22
Neptune	3.88	17.1	16.11 hrs	28.8	0.14
Pluto	0.24	0.002	6.39 days	50?	—

[a] Relative to Earth's

[b] With respect to the perpendicular to the orbital plane

[c] Maximum field strength in gauss at 1 radius from center of planet

Mars now have barely detectable magnetic fields. Earth is the exception among the terrestrial planets in having a moderate field.

In contrast, the icy giants possess fields equal to or stronger than Earth's (see table 7.1). Jupiter's surface field, for example, is fourteen times stronger and extends as a magnetosphere nearly to the orbit of Saturn. If we could see Jupiter's magnetosphere from Earth, it would appear as large as the full Moon.

Astronomers study the magnetospheres of other planets not only for their intrinsic interest but also for the opportunity to place Earth's system in a broader context. As we'll see, there are some interactions between moons and planets that we couldn't have guessed from looking at ours.

In this chapter we'll survey the magnetic fields and magnetospheres of the planets. Along the way, we'll see how studies of planetary magnetic fields offer clues about their internal structures. Let's begin with the weak sisters in the solar system, the Moon and the terrestrial planets.

THE SILVERY MOON

Everyone has heard about Neil Armstrong's grand declaration as he stepped out on the Moon for the first time, and most people know something of the discoveries the astronauts made afterward. The low gravity, the huge craters, and the black dust: all these made the science pages of the newspapers. But few people have ever heard about the Moon's magnetic field.

There's a good reason for that: the strongest field is a hundred times smaller than Earth's. The Moon has no dipole field, no magnetosphere, nothing similar to the Earth's field. What is there to read about?

It's not as if the astronauts didn't search thoroughly. They measured the field at four different landing sites during the Apollo 12, 14, 15, and 16 missions (1969 to 1972). They found a field that varies from place to place, with intensities ranging from a low of 6 nT to a high of 313 nT. (A nanotesla, nT, is 100,000 times smaller than a gauss.) For comparison, Earth's polar field is about 30,000 nT. These are minuscule fields indeed, locked in ancient magnetized rocks.

The astronauts brought samples of these rocks back to Earth. From the strength of the fields and the ages of the rocks, scientists concluded that the Moon probably had a magnetic dynamo operating in its core, between 3.6 and 3.9 billion years ago. After that relatively short period, the dynamo must have shut down. The question is "Why?" As we've learned in previous chapters, a dynamo similar to Earth's requires a partially liquid core that is spinning, as well as convection to convert toroidal to poloidal fields. Either the Moon's core solidified completely or the convection ceased or something else happened. Lunar geologists have debated the choices for decades. Happily, a few observations have helped to decide among them.

The Apollo astronauts placed seismometers at five different sites on the Moon. These instruments worked without a flaw for seven years, transmitting the faint tremors of "moonquakes." The strongest quakes only ranked a 2 on the Richter scale and were caused by the tidal action of the Earth on the Moon, and by meteor impacts. But the multiple recordings of each quake yielded precise information on how seismic waves propagate through the Moon. From such data lunar geologists constructed a tentative model of the Moon's interior. They determined that it has a small dense core, perhaps 800 km in diameter. They also guessed that the core consists of iron or an iron alloy.

A clever experiment aboard the Lunar Prospector produced more helpful data. The Prospector was a small spacecraft that orbited the Moon for eighteen months, beginning in January 1998. Its mission was to map the surface composition and to measure gravitational and magnetic fields. In April 1998 the Prospector and the Moon passed through the tail of the Earth's magnetosphere. Lon Hood and his colleagues at the University of Arizona measured a dip in the Earth's field strength. They attributed the dip to currents induced in the Moon by the Earth's field. And from the size of the dip, they concluded that the Moon has an electrically conducting core about 680 km in diameter. Their result still didn't show whether the Moon had a liquid or solid core, however.

A recent analysis of some very precise observations supports the idea of a liquid core. The Apollo astronauts also placed three "retroreflectors" on the Moon. Each one returns a beam of laser light directly back to its source. Beginning in the 1990s, powerful laser beams were flashed at these reflectors at regular intervals from Earth-based telescopes. From the measured round-trip travel time of the light, scientists could determine the distance to an individual reflector within a few centimeters, over a distance of 380,000 km (note 7.2).

With recent improvements in laser ranging, the uncertainty in a distance measurement has shrunk to a few millimeters. Such precision allows the wobble of the Moon's rotation to be clocked very accurately. In 2000, James Williams and his team at the Jet Propulsion Lab of Caltech drew several surprising conclusions from the data. The lunar core is *molten,* has a diameter of about 700 km, and is slightly elliptical. Moreover, it flexes under the tidal influence of the Earth, and as a result, it dissipates heat. Williams and friends suggest that this heat might have powered convection and a dynamo earlier in the Moon's history.

David Stevenson at Caltech disagrees. He is an expert in the subject of planetary dynamos who has been working in the field for twenty-five years. We'll be referring to his work constantly in this chapter. He estimates that the rate of heat loss from the core was never strong enough to drive convection. He proposes instead that the forth-and-back tipping of the Moon's axis (called nutation) was sufficient to stir the liquid core mechanically and drive a dynamo. In his scenario, the dynamo died when the core cooled sufficiently to become viscous.

David Stegman at UC Berkeley has yet another scenario. He and his international team proposed that convection was turned on briefly during an instability of the mantle. They pictured a dense "thermal blanket" at the base of the mantle that initially prevented the liquid core from cooling. Convection and magnetic field generation were therefore suppressed. Eventually, radioactive heating of the blanket increased its buoyancy so that it eventually rose back into the mantle. The removal of the blanket allowed the core to cool, convection developed, and a dynamo generated a magnetic field for a short period.

As you can see from these examples, there is still no consensus among theorists on exactly why the Moon's dynamo worked for a while and then died. Perhaps new data from the Lunar Ranging Program will help to sort out matters. In the meantime, analysis of older data continues to reveal surprises.

Here is one. Robert Lin and his team at UC Berkeley built a magnetometer for the Lunar Prospector, the spacecraft that was placed into orbit around the Moon in 1998. When Lin and company analyzed their magnetic data, they discovered that the strongest lunar fields lay diametrically opposite the largest impact craters, such as Mare Imbrium and Mare Serenitatis. Their result suggested that when a large meteor crashed into the Moon, billions of years ago, a doughnut-shaped cloud of ionized debris swept the existing magnetic field into a region directly opposite the point of impact. When the debris fell to ground, it trapped the field in a patch of magnetized rocks as large as several hundred kilometers in diameter. Here was additional proof that the Moon possessed a global field long ago.

MERCURY

The tiny planet nearest the Sun is a hellish place. Facing into the intense sunlight, its bare cratered surface swings from about 500 degrees Celsius (C) during the day (hot enough to melt zinc) and a nighttime temperature of -180 C. Mercury rotates very slowly; it takes fifty-nine Earth days to turn once. It is also the second smallest planet (only Pluto is smaller); its mass is only 5.5 percent of Earth's. The low surface gravity and high temperatures stripped the planet of any atmosphere it might have had.

What little we know about Mercury's magnetic field derives from the three flybys of the space probe Mariner 10, in 1974–1975. At altitudes of several hundred kilometers, a maximum field of 400 nT was measured, comparable to the surface field of the Moon. But to everyone's surprise, the shape of the field is consistent with a dipole. Moreover, the field lines point north, like Earth's.

Theorists are having trouble explaining Mercury's dipolar field. On the one hand, it is too strong to be just a remnant of an extinct dynamo, preserved in the surface rocks. On the other hand, it is too small to be the product of an Earth-like dynamo that is still working.

The key to the problem lies in the condition of the core. Mercury has the highest average density of any planet, which implies a metallic core as large as three-quarters of the diameter. Because Mercury is the least massive of all the planets, it must have cooled faster than Earth after forming 4.5 billion years ago. One might expect, therefore, that the core has solidified and that no dynamo could still be functioning. If some part of the core were still liquid, however, a dynamo unlike the Earth's could still exist.

Early models of the core by Swiss scientist P. E. Fricker and co-workers suggested that the outer core could still be molten. They took into account the heat released as the inner core solidifies (an effect similar to the freezing of ice) and the heating from the decay of radioactive elements. But their calculations were fraught with uncertainties about the chemical composition of the core.

David Stevenson and his colleagues also argued for a partially liquid core. They suggested in 1980 that a thin shell of liquid iron could have survived if a small amount of sulphur were present, raising the freezing point of pure iron. The upward migration of sulphur would then drive a dynamo, they claimed.

But if Mercury has a dynamo, it can't be similar to Earth's, in which cyclonic convection plays a key role. Even though Mercury rotates so slowly, the Coriolis effect would still be sufficient to create a dipole field thirty times too large. So theorists have explored two extreme possibilities to limit the efficiency of convection. Either Mercury's dynamo operates exclusively in a thin liquid shell or deep in a viscous core.

Sabine Stanley and her co-workers at Harvard University took the first position. In 2004 they used a three-dimensional numerical model of a dynamo to investigate whether a thin shell could generate a dipolar field that is also sufficiently weak. They found that convection can occur in a thin shell, but just barely. If the shell were also viscous enough, it could indeed produce the feeble fields that are observed. (We'll hear much more about Stanley and her work later on.)

Ulrich Christensen at the Max Planck Institute for Solar System Research, in Lindau, Germany, examined the opposite point of view in 2006. He pictures a liquid shell in which convection is completely absent and serves only to filter and weaken the field. He demonstrated that a dynamo could operate below the shell in the semiliquid top of a solid core. Mercury's slow rotation is critical in this dynamo. It generates a complicated multipolar field with dipolar, quadrupolar, and octupolar components. In Christensen's model the minor components dissipate their energy as heat as they diffuse through the stable liquid shell. Only a weak dipole field manages to survive to reach the surface.

Three geologists at Caltech have recently challenged the thin shell model. They claim that if the core were almost entirely solidified, as the shell model assumes, then the planet would have contracted more than certain crustal features allow. So they recalculated a variety of thermal histories of Mercury in an attempt to understand its puzzling field.

Jean-Pierre Williams, Oded Aharonson, and Francis Nimmo varied the initial core temperature and the percentages of sulphur and potassium. Only a few extreme combinations of trace elements resulted in a present-day dynamo. In most of their trials, the core is still partially molten, even after 4.5 billion years, but doesn't produce a lasting dynamo. These results are controversial and will be challenged in turn, we may expect.

Way back in 1987, David Stevenson proposed an entirely different form of dynamo. He showed that the thermoelectric, or electrochemical, effect at the core-mantle boundary could generate electric currents and, therefore, toroidal fields (note 7.3). He postulated that horizontal temperature gradients in the planet's mantle might convert the toroidal field to a poloidal field. Two scientists at the Imperial College of London have picked up the idea recently and find that it has merit.

At present no explanation for Mercury's field stands out as compelling. The problem has remained unsolved for thirty years, in part for lack of information. In fact less is known about Mercury than any other planet, with the possible exception of Pluto. For this reason NASA decided in the late 1990s to send an orbiting observatory called MESSENGER to Mercury. (As you may remember, the god Mercury was messenger to the more senior Roman gods. He wore tiny wings on his feet to speed him on his errands.)

MESSENGER was launched on August 3, 2004, and will go into orbit around Mercury on March 18, 2011, if all goes well. Its prime mission is to explore Mercury's tiny magnetosphere. Instruments onboard, designed by scientists at Johns Hopkins University, will sample the magnetic field, the energetic particles, and the plasma for about a year. Until then we'll have to wait to test the menu of competing models.

THE GODDESS OF LOVE AND BEAUTY

Venus is similar to Earth in size, composition, and age, but there the similarities end. The planet has a noxious atmosphere of carbon dioxide, nitrogen, and sulphuric acid, ninety times as dense as ours. This heavy blanket of gases acts as a greenhouse, trapping sunlight and maintaining the surface at a blistering 500 C. The rocky surface has been explored with Earth-based radars, orbiting spacecraft, and several Soviet landers. It is a barren landscape of craters, ancient volcanoes, and vast lava fields. The planet also has a peculiar motion, rotating in 243 Earth days and orbiting in 225 days. Thus, the planet rotates very slowly and in the opposite direction as Earth and most other planets.

In 1962, the Mariner 2 space probe flew past Venus and failed to find any sign of a magnetosphere. A series of Soviet and American satellites in the 1970s gradually whittled down the upper limits of the planet's field strength. Finally, in 1979 the Pioneer Venus Orbiter established that the field was at least 100,000 times weaker than Earth's.

So here is a puzzle. How can tiny Mercury maintain a weak dipole field and a magnetosphere, when the more massive Venus cannot? Is it possible that Venus has a completely solid core, while Mercury has retained a thin liquid shell? Or is there some other explanation for Venus's lack of a field?

David Stevenson and colleagues carried out extensive modeling of the terrestrial planets in 1983 and came up with a surprising conclusion: the core of Venus could be completely molten. Since they assumed that the core and mantle of Venus has the same composition as Earth's, what accounted for the difference? They calculated that Venus's hot atmosphere acts as a thermal blanket, preventing the core from solidifying. Without the additional heat released by the freezing of the core, convection currents would fail to develop. Without convection, a dynamo similar to Earth's would not be viable.

Francis Nimmo, a geologist at the University of London, has recently offered a different explanation for the lack of convection in Venus's core. He pointed out that plate tectonics, the slow movement of giant segments of the Earth's upper mantle, drains energy from the core. The energy is transported from core to mantle by convection currents, which are essential parts of Earth's dynamo. In contrast, many surveys of Venus's surface show no signs of plate tectonics for at least the past 500 million years. As a result, Nimmo hypothesizes that Venus's dynamo shut down when plate tectonics ceased. He allows that Venus might have had a working dynamo before then.

Although Venus doesn't generate a magnetic field now, it does have a magnetosphere of sorts, which may seem contradictory. The Pioneer Venus Orbiter (1979–1981) revealed that the solar wind interacts with a conducting layer of ions at the top of the atmosphere. This so-called ionosphere owes its existence to the intense solar radiation that strips atoms of their electrons. The layer is electrically conducting and acts as a barrier to the interplanetary field.

Christopher Russell and Janet Luhmann, two of the Orbiter scientists at UCLA, have described Venus's "induced" magnetosphere. It has neither a tail nor belts analogous to Earth's Van Allen belts. In fact the whole structure is extremely compact.

Ahead of the planet, upstream in the wind, lies a bow shock. The wind and its embedded field slow down as they pass through the shock and wrap around the planet as a "magnetosheath." The wind plasma compresses the day side ionosphere until they reach the same pressure in a surface called the ionopause, at an average height of about 300 km. Some interplanetary field leaks through the ionopause in the form of twisted magnetic ropes, with strengths of about 100 nT. In addition a weak magnetic field is induced in the ionosphere by the fluctuating pressure of the solar wind. The interactions are complex, however, and we won't venture further with them. Suffice to say that Venus's magnetosphere is a combination of induced and interplanetary fields.

Venus also shimmers with dim auroras. A spectrometer aboard the Orbiter detected a spectral line at a wavelength of 130.4 nm, which is emitted by atomic oxygen. Venus's auroras are unlike the spectacular displays on Earth, because Venus has no magnetic field to guide fast wind particles to the night side. Instead, the Venusian auroras occur at the dawn and dusk sides of the day side of the planet. They are sporadic and vary in brightness according to the solar wind speed. Evidently they are caused by impacts of electrons with energy of about 300 volts that collide with oxygen atoms. As we'll see, the giant planets also generate auroras.

We'll soon learn more about Venus's atmosphere and magnetosphere. On November 9, 2005, the Russians launched the Venus Express, a European spacecraft, from the Baikonur Cosmodrome in Kazakhstan. This vehicle carries an armory of instruments, including a sensitive magnetometer designed by a team in Austria. The spacecraft went into orbit around Venus on April 11, 2006, and is slated to observe for at least a year. As of this writing, the instruments are performing well. A few preliminary results were published in 2007.

THE RED PLANET

Mars has fascinated the public and professional astronomers for more than a century as the planet most likely to harbor life. Italian astronomer Giovanni Schiaparelli saw linear markings on the planet in 1877 and called them *canali,* the Italian for channels. The word was translated as "canals" in English and set off a storm of speculation about their origin. Like many educated Bostonians of his time, Percival Lowell was convinced that they were evidence of an advanced civilization. He established a personal observatory in Flagstaff, Arizona, to confirm his belief.

Alas, he was mistaken (note 7.4). But as I write these lines, two robotic rovers (Opportunity and Spirit) are cruising the red planet in search of water, and hopefully, signs of ancient life. In 2003 a trace of methane was discovered in the thin carbon dioxide atmosphere, with the tantalizing possibility that some form of life still produces it. Moreover, in March 2007, NASA announced that a tremendous volume of water ice covers the poles, which is also an encouraging sign.

Only the simplest forms of life could survive on Mars, though. Its surface temperature varies from 20 C to -140 C, and howling dust storms rage over the planet for weeks. There is no liquid water, though there used to be. Oxygen is present in the atmosphere only at one part in a thousand. Mars lacks all the amenities.

Mars also lacks an Earth-like magnetic field. Five Soviet MARS satellites, launched between 1971 and 1974, couldn't detect any field, but they did prove that Mars has an ionosphere that deflects the solar wind. So in principle, Mars could have a weak induced magnetosphere similar to that of Venus. Then in 1989 the Soviet spacecraft Phobos 2 determined that the strongest field in the wake of Mars was at least 10,000 times weaker than Earth's.

NASA's Mars Global Surveyor was launched in November 1996 and has provided the most detailed information yet on the magnetic field and the interior of Mars (note 7.5). The satellite was intended to observe for only two years but functioned beautifully for ten. In November 2006, however, a computer glitch resulted in a fatal maneuver. Nevertheless, the mountain of data the craft spat out will keep scientists busy for a long while. Some exciting results are being published now.

The Surveyor circled Mars in a polar orbit every two hours, scanning the entire surface from an altitude of 400 km. An onboard battery of instruments included a laser altimeter, a high-resolution camera, and a magnetometer. Mario Acuña at NASA's Goddard Space Flight Center leads the magnetometer team. In 1999, he announced the results from the first 6,000 orbits around the planet. Mars has no global field now, he said, but does have a strongly magnetized crust. Consequently, Mars must have had a dynamo in the past.

From the absence of a magnetized crust near ancient craters, his team was able to determine approximately when the dynamo died. When asteroid or comet impacts created these craters, they also erased part of the crust's magnetism. The craters were never remagnetized. Because these demagnetized regions are some 4 billion years old, the dynamo must have expired at least that long ago. That allows only about half a billion years for the dynamo to have magnetized the crust.

The most exciting results from the magnetometer concern the magnetic pattern in the crust. Jack Connerney, a member of the team, reported in 1999 that they had compiled a complete magnetic map of the planet from measurements at an altitude of 400 km. The map (fig. 7.1) shows that the crust is magnetized in stripes several hundred kilometers long that are concentrated almost entirely in Mars's southern hemisphere. The crustal fields are

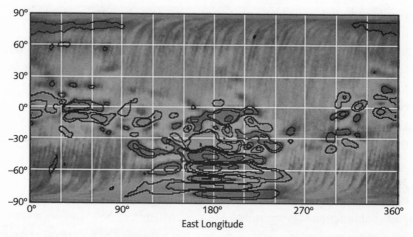

Fig. 7.1. Maps of the magnetic field of Mars at 400 km altitude, obtained by the Mars Global Surveyor. The crustal field alternates in sign like the seafloor of the Earth, but the cause might be quite different.

at least a thousand times stronger than similar fields on Earth. Moreover, the stripes alternate in magnetic polarity, like the stripes on Earth's seafloor (note 7.6).

One explanation for the stripes, therefore, is that the ancient dipole field must have flipped over periodically, exchanging north and south poles, as on the Earth. As Connerney said in a 1999 press conference, "If the bands on Mars are an imprint of crustal spreading, they are a relic of an early era of plate tectonics on Mars. However, unlike on Earth, the implied plate tectonic activity on Mars is most likely extinct."

Connerney's teammate Mario Acuña cautioned that an alternate interpretation is possible. Cracking and shifting of an ancient, uniformly magnetized crust, due to volcanic activity or tectonic stresses, might have produced the stripes. "Imagine a thin coat of dried paint on a balloon, where the paint is the crust of Mars," he said. "If we inflate the balloon further, cracks can develop in the paint, and the edges of the cracks will automatically have opposite polarities, because nature does not allow there to be a positive pole without a negative counterpart."

Mars's crust is not the only site of exciting discoveries. Two scientists at Caltech's Jet Propulsion Laboratory have teased out crucial information about Mars's core, the location of the ancient dynamo. They tracked the variations in the frequency of a radio beacon on the Global Surveyor, which are caused by the Doppler effect. From these data they determined that the spacecraft's orbit is influenced by a tiny bulge (less than 1 cm) in Mars's equator, caused by the tidal pull of the Sun.

Charles Yoder, one of the scientists, explained in 2003 that by measuring this bulge, they could determine how flexible the body of Mars is. The results showed that Mars's iron core couldn't be entirely solid; it may either be wholly liquid or consist of a solid center and a substantial liquid shell. They also estimated the size of the core as about half the size of the planet.

Theorists and modelers have leaped on the flood of data from the Mars Global Surveyor and are scrambling to interpret the main results. We'll focus here on attempts to reconstruct the thermal and magnetic history of the planet. As of this date the debate is rapidly evolving, with many competing proposals on the table.

There seems to be general agreement on a few points. First, Mars had an Earth-like dynamo for no more than half a billion years after it formed. Fields as strong as 400 nT were frozen into the crust during that time. Second, such a dynamo requires convection currents in a partially or wholly liquid core. Third, convection only exists if a critical temperature gradient from the core through the mantle is exceeded. So, the core and mantle must be understood as a coupled system.

The key question facing theorists is "Why did the dynamo shut down when it did?" Something turned off the critical convection currents, and researchers have offered several ideas on how that might have happened. Here is a brief sample.

In David Stevenson's early scheme, the iron core remained completely liquid because of a high content of sulphur. Strong convection would support a dynamo for a while but would cool the core so quickly that ordinary heat conduction could take over as a means of transporting energy to the mantle. The convection would die and, with it, the dynamo.

Stevenson and Francis Nimmo came up with another idea in 2000. They hypothesized that plate tectonics of the Martian surface was involved in the demise of the dynamo. In tectonics, cold plates near the top of the mantle dive into the warmer depths and promote heat transfer from the core. If plate tectonics were to stop, say, by the freezing of all the plates into one huge plate, the heat loss of the core would slow, convection would turn off, and the dynamo would be stifled. Their models seemed to bear out this scenario.

But why would tectonics stop? Martin Collier and Adrian Lenardic at Rice University teamed up with Nimmo in 2004 to explain. Their model showed that the growth of a thick buoyant crust could insulate the lower mantle and squelch convection within it.

Nimmo and Jean-Paul Williams at UCLA recognized later that observational evidence for plate tectonics on Mars was really rather weak. So instead of cooling the mantle with tectonics to drive convection, they raised the initial temperature of the core. Their models show that if a wholly liquid core were only 150 K hotter than the mantle when the planet formed, convection and a dynamo would start. As this temperature excess drained away, the dynamo died. A theory like this, which depends on juggling the initial conditions, is subject to criticism, however.

In an alternate scheme, the core lacks sulphur and therefore solidifies rapidly, leaving a liquid shell. The dynamo operates in the shell as the inner core continues to grow, until the shell becomes so thin that convection currents are stifled. At that point, presumably after half a billion years, the dynamo dies.

You can see from this brief sample that considerable uncertainty remains about the early thermal history of the planet. The duration of the ancient dynamo depends sensitively on trace elements like sulphur and potassium in the core and the physics of slow convection in a semisolid mantle. It is also possible that surface effects squelched the dynamo.

Those intriguing magnetic stripes have also received a lot of attention. As we saw earlier, Jack Connerney and his colleagues view the stripes as similar to seafloor spreading on Earth. In their view the crust was spreading while Mars's magnetic dipole flipped from north to south several times.

Kenneth Sprenke and Leslie Baker at the University of Idaho have examined this idea, and it seems to hold water. They can match the observed polarity and positions of the stripes if the spreading center were far to the north of the stripes. (They didn't explain why the dipole flipped, however.)

A pair of scientists at McGill University in Montreal located the poles of the dipole in this scenario from a careful dating and matching of magnetic patches. They determined that the north and south poles wandered independently over a whole quadrant of Mars's surface. Earth's magnetic poles also wander, as we saw in chapter 2.

In summary this is a time of great ferment and excitement among planetary astronomers and geologists. The data are rich and the theories are flying thick and fast. We'll wait to see how it all shakes out.

BY JOVE!

The ancient Romans named the planet Jupiter after the supreme god in their religion. They knew it well as the fourth brightest object in the sky. Nothing more was learned about the planet until 1610, when Galileo, that incomparable astronomer, turned his tiny telescope to it. He was the first to see the four largest moons (Io, Europa, Ganymede, and Callisto). The view of this miniature planetary system convinced him that Copernicus's vision of our solar system was correct. In July 1664 Italian astronomer Giovanni Cassini, for whom NASA named a spacecraft, discovered the Great Red Spot, a rotating storm in Jupiter's atmosphere three times the diameter of Earth. This monster has persisted for the past four hundred years, feeding on the heat that emerges from the planet.

Jupiter has been a prime target of astronomers ever since these early years. Its turbulent atmosphere has been studied with ground-based telescopes and by the Hubble Space Telescope. It's also been probed by the spacecrafts Pioneer 10 and 11 in 1973; Voyagers 1 and 2 in 1979; Ulysses (1990–2009); Galileo (1989–2003); and currently by Cassini.

In the winter of 2000, Cassini was flying past Jupiter on its way to Saturn. At a distance of 10 million kilometers, a special camera onboard, named MIMI, took snapshots of a gigantic balloon that surrounds Jupiter. This was Jupiter's magnetosphere, imaged in a way that no other spacecraft had been able to do. It is the largest object in the solar system, with a tail that stretches to the orbit of Saturn. If we on Earth could see this monster, it would appear as large as the full Moon.

In this section, we'll talk about Jupiter's magnetic field and a few of the odd things scientists have learned about it. We'll also see how Io, one of Jupiter's largest moons, interacts with the field and inflates it with charged particles. But first, a little background.

Jupiter Revealed

Unlike the terrestrial planets, Jupiter is composed primarily of light elements like hydrogen and helium, 75 and 24 percent by mass, respectively, with a sprinkling of other gases like carbon dioxide, ammonia, and methane. At five times the Earth's distance from the Sun, Jupiter receives so little heat that the temperature at the top of its atmosphere falls to -150 C.

The giant planet is a collection of superlatives. It is eleven times larger in size than Earth and three hundred times larger in mass; it has sixty-three moons at last count; and has the shortest day (9 hours, 55 minutes) of any planet. Winds near its equator blow as fast as 360 km/hr in a series of colored bands parallel to its equator.

Spectroscopists tell us that the planet's atmosphere is made up primarily of gaseous molecular hydrogen, with traces of carbon dioxide, ammonia, water, and methane. The cloud tops are sprinkled with ammonia ice crystals. We don't really know how deep the atmosphere is, but the Galileo probe found no solid surface down to 150 km. In fact, Jupiter probably has no solid surface. What lies below? We have to rely on theoretical models to get some idea.

French astronomer Tristan Guillot and his co-workers at the University of Nice have recently reviewed what we think we know about the structure of the interior. In the simplest models of the interior, a dense core of 10 to 15 Earth masses and of uncertain composition lies at the center. Overlying the core and extending to at least three-quarters of the radius is an envelope composed of a weird substance called metallic liquid hydrogen.

In 1935, Cornell University physicist Eugene Wigner predicted that such a material could exist. He calculated that under the extreme pressures that must prevail in Jupiter's interior (as high as 40,000 times the atmospheric pressure on Earth's surface), hydrogen gas becomes a liquid with the electrical properties of a metal. This is an exotic state of matter in which electrons are detached from the atoms and are able to move freely. Wigner predicted that the interior would therefore be a good conductor of electricity.

Metallic hydrogen in the envelope changes into liquid molecular hydrogen where the internal pressure drops to about 2,000 atmospheres. At lower pressures higher up, liquid hydrogen becomes gaseous and extends outward to form the atmosphere.

Infrared observations of Jupiter show that it emits more heat than it receives from the Sun. The surplus of energy could come from several sources;
the most likely is the slow contraction of the envelope, with the release of
gravitational energy. As a result of contraction, the core temperature could be
as high as 20,000 K, three times the surface temperature of the Sun. The temperature decreases outward to -150 C at the top of the atmosphere. Such a
strong temperature gradient could drive convection currents in the liquid
hydrogen envelope and atmosphere. As we know, these currents would be
essential for an internal dynamo.

Jupiter's Magnetic Field

Astronomers might have guessed that Jupiter has a magnetic field, but they
had no experimental evidence until 1955. In that year Bernard Burke and
Kenneth Franklin, two radio astronomers at the Carnegie Institution of Washington, discovered that Jupiter emits powerful bursts at a wavelength of 13.5
meters. At first they thought the radiation was produced by thunderstorms
in Jupiter's atmosphere. When they determined later that the radiation was
elliptically polarized, though, they concluded that it must originate in a magnetic field. Theirs was the first detection of a planetary field other than the
Earth's. Four years later, radio astronomer Frank Drake obtained a spectrum
of the bursts that proved they were nonthermal emissions, probably due to
high-energy electrons in a strong magnetic field.

Pioneer 11 was the first spacecraft to probe Jupiter's field. In 1974 it flew
within 1.6 Jovian radii from the center of the planet and detected a field fourteen times stronger than Earth's. Edward Smith, a veteran planetary astronomer at NASA's Jet Propulsion Laboratory, was the leader of the team that built
the magnetometer aboard Pioneer. When the spacecraft passed safely through
Jupiter's powerful radiation belts, similar to Earth's Van Allen belts, Smith
and his crew were ecstatic.

Their magnetometer went on to map enough of Jupiter's magnetosphere
for Smith and company to present a first picture of its shape. Within 20 Jovian radii of the planet, the field is a dipole, at least ten times stronger than
Earth's. It has the opposite polarity to Earth's, with field lines pointing from
north to south, and rotates at the same rate as the planet. In the middle distance (20 to 60 radii) the field is distorted by trapped electrically charged

particles and rotates more slowly than the planet. In this region, ionized particles circle in the planet in a current ring, which generates a secondary magnetic field that is stronger than the dipole field at large distances. Beyond 60 radii, a magnetotail stretches far out into space away from the Sun.

Irene Engle, a physicist at the U.S. Naval Academy, combined data from the Pioneer and Voyager flybys to compute a three-dimensional model of the field (fig. 7.2 top). The scale of the field is enormous: the day side lobe extends about 100 Jovian radii into the solar wind, and the tail stretches 4.5 astronomical units (AU) to the orbit of Saturn. (An astronomical unit is the distance from Earth to Sun, 8 light-minutes, or 140 million kilometers.) The field is also more complicated than Earth's, with quadrupole and octupole components much stronger relative to the main dipole field. This result would have implications concerning the site of an internal dynamo.

The most puzzling aspect of the magnetosphere was the so-called magnetodisk. Why was the middle magnetosphere between 20 and 60 radii so much flatter than Earth's? The answer to this question turned out to involve a complex interaction between Io and Jupiter. Voyager 1 revealed some clues in 1979, and Galileo confirmed them in 1995.

Voyager 1 was an amazingly productive spacecraft. It was built to tour all the giant planets and the regions beyond. It was launched in 1977, and thirty years later it is still working, sending data from 100 AU, beyond the terminal shock of the heliosphere. When it passed Jupiter in March 1979, the planet relayed over 30,000 superb images of the planet and its moons. Among the many discoveries this durable satellite has made, the volcanism of Io stands out.

Io, it turned out, has at least one hundred active volcanoes, more than any other body in the solar system. They are constantly erupting, spewing sulphur dioxide gas and sulphur particles hundreds of kilometers into space. (A camera on the Hubble Space Telescope saw one rise 400 km above the surface.) They maintain a thin atmosphere of sulphur dioxide around the little moon. Planetary astronomers attribute the volcanic activity to the solid tides (100 m high) that the gravity of Jupiter raises in Io's surface. Io flexes under these forces, which heat its interior and force lava to the surface.

Io's volcanism would be interesting but irrelevant to Jupiter's magnetic field except for the fact that its orbit, at 6 radii, lies entirely within the belt of

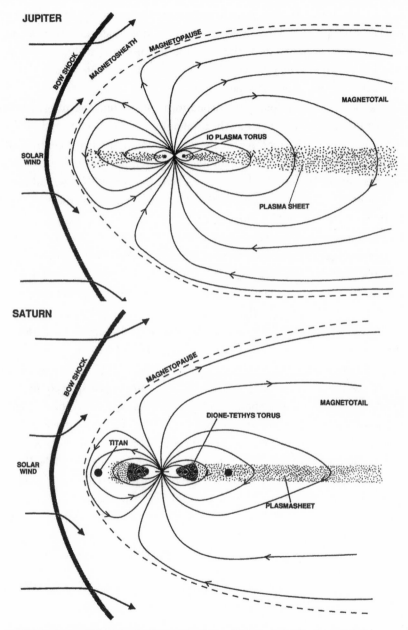

Fig. 7.2. Cartoons of the magnetospheres of Jupiter and Saturn. Jupiter has a pronounced flattened magnetodisk, as compared with Saturn.

energetic trapped charged particles. These particles bombard Io's sulphur dioxide atmosphere and kick atoms into orbit around Jupiter at a rate of about 1 ton per second. At this point they are orbiting the planet at about the same speed as Io, while several processes ionize them. Then Jupiter's rotating dipole field sweeps by and accelerates them.

The net result is a doughnut-shaped torus of sulphur and oxygen ions that *co-rotates* with Jupiter along the orbit of Io, at a distance of 6 radii from the center (see fig. 7.2 top). As the mass in the torus builds up, centrifugal forces cause it to expand radially outward, stretching lines of force of the planet's dipole. As a result, the field lines in the range of 20 to 60 radii are pulled into the shape of a flattened disk that contains a plasma sheet.

In 1998, Debbie Huddleston, Christopher Russell, and their colleagues at UCLA showed how the balance of magnetic, gravitational, plasma pressure, and centrifugal forces shape the magnetodisk. In the direction toward the Sun, the disk's expansion is stopped by the pressure of the solar wind. In the tail direction, however, the field beyond about 23 radii is too weak to restrain the pressure of Io's ions, and they stream outward in a flat plasma sheet. Eventually, these ions pick up electrons and become neutral atoms that can escape from the magnetosphere. It is this cloud of atoms that MIMI, the special camera on Cassini, photographed.

But Io's influence on its parent planet doesn't stop there. Io is connected to Jupiter's north and south poles by magnetic field lines. As Io orbits around Jupiter and through the plasma torus, a powerful electric field drives a current of sulphur and oxygen ions along these field lines. The accelerated ions smash into Jupiter's poles and produce auroras that glow in ultraviolet light. John Clarke at the University of Michigan obtained sharp images of the aurora using the Hubble Space Telescope in October 1996. The images show that the Jovian auroras are a thousand times brighter than those on Earth. In combination with observations from Galileo, the images helped to unravel the intricate connections between Io and the magnetosphere.

Europa

We've left the best for the last. In 1997 the Galileo orbiter completed its first magnetic survey of the four Galilean moons. Each one was different. Callisto had a minuscule field, no stronger than 30 nT. Io and Ganymede both had significant dipole fields, however, with strengths of 2,600 nT and 1,500 nT, respectively. The dipoles were nearly aligned with the spin axes. Margaret Kivelson, principal investigator for the UCLA magnetometer, thought they could be generated by dynamos.

Europa's field was weird. The best way to describe it was a dipole with a strong quadrupolar component tilted 65 degrees from the dipole axis. If future observations confirm that interpretation, Kivelson said, it could mean that the field is generated in a shallow shell near the surface.

There were other data that suggested a shallow layer under the surface. Galileo (the satellite, not the probe) had obtained images of Europa's surface that showed a cracked icy layer, resembling a field of ice floes floating on a liquid ocean. Scientists were fascinated by the possibility of a hidden ocean, just below Europa's surface. If that proved to be true, it would be the only place in the solar system other than Earth where liquid water exists. Could such a hidden ocean correspond to the shell Kivelson had talked about?

Galileo revisited Europa in January 2000 and caused a sensation. Margaret Kivelson presented the new findings. "Jupiter's magnetic field at Europa's position changes direction every 5-½ hours," she said. "This changing magnetic field can drive electrical currents in a conductor, such as an ocean. Those currents would produce a field similar to Earth's magnetic field, but with its magnetic north pole—the location toward which a compass on Europa would point—near Europa's equator and constantly moving. In fact, Europa's field is actually reversing direction entirely every 5-½ hours" (M. Kivelson et al., *Science* 276, 1239–41, 2001).

Here was convincing evidence for a conductor, possibly an ocean of salty water, deep below the icy surface of Europa. Astrobiologists were excited at the prospect that life of some form might exist in such an ocean. Conferences have sprouted up ever since to debate the possibilities. At the same time, lunar geologists are working to fit an ocean into a model of the structure and evolution of the little moon.

Origin of the Field

We turn now to the source of Jupiter's primary dipole field. The planet would seem to have all the requirements for an internal dynamo: a fast spin, an electrically conducting liquid interior, and quite possibly, convection currents that carry heat from the interior to the surface. If we had complete confidence in our models of the interior, we could compute the output of a dynamo and compare it to the observed field. Or if we had a way to extrapolate the observed field into the interior, we could check our models of the structure of the interior. But, as Tristan Guillot and his colleagues have written, neither route to a complete picture is without uncertainties.

The best they were able to do was to investigate whether a reasonable model of the interior could predict certain critical parameters of a dynamo. They checked the order of magnitude of the dipole field, the ratio of quadrupole to dipole components, and whether convection would move field lines faster than they would otherwise diffuse through the conducting metallic hydrogen. In all cases the tests proved encouraging. An Earth-like dynamo is likely, if not certain.

A few intrepid theorists have attempted to simulate a Jovian dynamo in a computer. Sergei Starchenko and Christopher Jones at the University of Exeter, for instance, have estimated convective velocities and field strengths for Jupiter and other planets, with some success. Gerald Schubert and his Chinese co-workers at the University of Exeter have shown that electrically conducting surface flows can strongly modify the output of a dynamo that operates at greater depths. Some evidence suggests that Jupiter's banded wind structure may extend as deep as a few thousand kilometers. Could these winds affect the dynamo?

Jupiter's dynamo, however it is constructed, must depend on the properties of liquid metallic hydrogen, which fills most of the interior. Until recently we had only theoretical calculations of its properties. Now several groups have undertaken experiments to measure them.

For example, at the Lawrence Livermore National Laboratory in Berkeley, California, William Nellis and his team have produced metallic hydrogen by firing a supersonic jet of gas at a sample of cold liquid hydrogen. The impact creates a shock wave in the sample, raising its pressure to several thousand atmospheres and its temperature as high as 4,000 K for a few microseconds. In that brief time, the researchers measured the pressure at which hydrogen becomes metallic and also the electrical conductivity of that state.

Their results suggest that Jupiter's magnetic field is not generated deep in the planet, as some scientists have assumed, but much closer to the surface.

Metallic hydrogen, with its high electrical conductivity, forms at a minimum pressure of 1,400 atmospheres. In the tentative model of Tristan Guillot and his co-workers, that pressure is reached at a depth of only one-tenth of the Jovian radius. The implications of these results remain to be worked out.

SATURN

Saturn's rings are among the first objects that children want to see through a telescope. They are beautiful indeed and deserve all the attention they get from the public, as well as from professional astronomers. But if we could see Saturn's magnetic field, as have Tom Krimigis and his team at Johns Hopkins University's Applied Physics Laboratory, we might be equally impressed. If you recall, the team used MIMI, a special camera aboard the Cassini satellite, to obtain an image of Jupiter's magnetosphere. They had another chance in June 2004 to do the same for Saturn from a distance of 6 million kilometers, or about 100 "Kronian" radii. Their image shows a cloud of neutral hydrogen atoms that have barely escaped from the magnetosphere. The cloud is enormous, 2.4 million kilometers in diameter. If we could see it from Earth, it would be about a fifth of the size of the full Moon.

Cassini-Huygens is the first satellite to visit Saturn since Voyager 1 and 2 in 1980–1981. It consists of an orbiting platform for instruments (Cassini) and a probe (Huygens) that descended into the Kronian atmosphere. Cassini circles the planet every seven days and is generating a flood of data. Its instruments have led scientists to some surprising discoveries and have helped to refine concepts derived from past flights. With Cassini's new observations, planetary astronomers have obtained some interesting insights into the interactions among a planet, its moons, and the solar wind, as we shall see shortly.

Saturn without its rings would look like a slightly smaller version of Jupiter. It's a flattened ball of frozen hydrogen and helium gases, ten times the diameter of Earth, with a dense core and a liquid metallic hydrogen envelope. Despite its size, Saturn's low average density would allow it to float in water if we could find a big enough bathtub. Like Jupiter it rotates rapidly (in about 10 hours, 40 minutes) and has a banded atmosphere with winds as high as 1,800 km/hr. Saturn also has many moons, thirty-three of them named, with another twenty-three waiting for attention.

Saturn has seasons like the Earth because its axis is tipped 29 degrees with respect to its orbit. But Saturn takes 29 years to circle the Sun, so a winter there would last a full 7.5 years. Not a pleasant prospect!

Saturn's Magnetosphere

Before Cassini arrived at Saturn, everything we knew about its magnetic field was discovered by the flybys of the Pioneer and Voyager spacecraft. Edward Smith, leader of Pioneer's magnetometer team, described its huge magnetosphere in 1981. It had a bow shock 24 Kronian radii upstream, a magnetopause at 17 radii, and an enormous magnetotail (fig. 7.2 bottom). Despite its size, Saturn's magnetosphere was still only about one-fifth as large Jupiter's. It was also a simpler system than Jupiter's, almost a pure dipole without a quadrupole component. And unlike any other planet, Saturn's rotational and magnetic axes were aligned to within 1 degree, a circumstance that might be relevant for the internal dynamo.

Saturn's dipole field was opposite in polarity to Earth's, with an equatorial strength of 0.3 gauss at 1 radius from the center. (Saturn's radius equals about 60,000 km.) That was twenty times weaker than Jupiter's and comparable to Earth's. Nevertheless, Saturn's magnetosphere fluctuated less than Jupiter's to changes in solar wind pressure. In this respect its behavior was closer to Earth's.

James Van Allen, one of the Pioneer 11 investigators, wrote that "in terms of planetary radii, the scale of Saturn's magnetosphere more nearly resembles that of Earth and there is much less inflation by entrapped plasma than in the case at Jupiter." He viewed Saturn as an intermediate case between Earth and Jupiter. That was an important insight for the time, but it is being challenged now.

Michele Dougherty and her team at the Imperial College of London are among the challengers. They built the magnetometer aboard Cassini and have been observing the magnetosphere for three years. Their observations suggest that, contrary to Van Allen's conclusion, Saturn's field more nearly resembles Jupiter's than Earth's.

The key issue is whether Saturn has a magnetodisk and, if so, how it behaves in the varying pressure of the solar wind. Earlier we described Jupiter's magnetodisk. Jupiter's rapid rotation creates centrifugal forces on Io's torus of cold ions. The dipole field is stretched into an extended flat disk. The field is weak in the direction of the Sun and responds violently to changes of solar wind pressure. Saturn also has sources of heavy ions and rotates rapidly, so one might expect it to develop a magnetodisk as well. But Voyager showed none.

Observations from Cassini now support the existence of a disk. Chris Arridge, a member of the magnetometer team, used 2.5 years of data to show that the same forces are acting in Saturn's field as in Jupiter's. He argues that the solar wind prevents the formation of a disk on the day side of Saturn. But on the night side, the field beyond 15 Kronian radii stretches out into the tail, forming a thin current sheet. The cause of the stretching is centrifugal force, as in Jupiter's magnetodisk.

Arridge and friends were able to model the field and plasma forces in the sheet and to demonstrate the similarity with Jupiter's disk. Dougherty added several other characteristics of Saturn's magnetosphere that are similar to Jupiter's, including reconnection in the tail and the motion of flux tubes. Altogether, these researchers believe that Saturn's field resembles Jupiter's far more than was earlier thought.

But other data conflict with this conclusion. For example, Cassini observations of Saturn's aurora show that some effects resemble Earth's, others Jupiter's. It is a complicated world.

Inside the Bubble

Thirty years of ground-based observations, combined with the early spacecraft data, gave us a fairly clear picture of the internal structure of Saturn's magnetosphere. The planet is surrounded by a series of doughnut-shaped rings of hot and cold atoms and ions, from a variety of sources (fig. 7.2 bottom).

Inside a distance from the planet's center of about 7 radii lies a torus of trapped high-energy electrons and ions. It's the analogue of Earth's Van Allen belts. A ring current associated with these belts lies between 8 and 16 radii and varies widely in position and width as the pressure of the solar wind varies.

Five small, icy moons orbit the planet between 3 and 10 radii. Dione at 5 radii and Tethys at 6 radii are the largest. They are composed of rock-hard water ice. Meteors and fast ions bombard their surfaces and produce low-energy protons, oxygen ions, and hydroxyl ions. These particles fill a thick ring, or torus, that encloses the moons and extends to greater distances as a thin plasma sheet in the equatorial plane. This whole system has been named the E ring.

One of Cassini's triumphs was the discovery that Enceladus, a tiny moon only 500 km in diameter, is the main source of plasma for the E ring. The moon orbits the planet at about 4 radii. Its icy surface reflects light almost perfectly, so its color is snow white. During a flyby in February 2005, the magnetometer aboard Cassini detected a distortion of Saturn's magnetic field near the moon. Michele Dougherty and her magnetometer team reported this discovery at a NASA press conference. "These new results from Cassini may be the first evidence of gases originating either from the surface or possibly from the interior of Enceladus," she said.

If so, the gases had to be ejected recently, because the moon's gravity is too weak to retain an atmosphere. That guess was confirmed in November 2005, when Cassini's infrared camera captured an image of a plume, or geyser, erupting from the south pole of the moon. The spectra of the plume indicated that it consisted of micron-sized water ice. Scientists now hypothesize that recurring geysers of ice maintain the density of the E ring and that Saturn's tidal forces on the moon provide the necessary energy.

Titan, Saturn's largest moon, also plays a star role in Saturn's magnetosphere. Titan is the second largest moon in the solar system, a notch below Jupiter's Ganymede and larger than Mercury. It is also the only moon with a dense smoggy atmosphere, which is composed mostly of molecular nitrogen, with traces of methane, ammonia, and water.

In December 2004, the Huygens probe was released from the Cassini orbiter and plunged at high speed into Titan's atmosphere, relaying vital information on the way down. It landed on a firm surface of icy boulders. Images taken during descent show large, dark areas that were thought to be liquid. In 2007, Cassini's radar penetrated the thick atmosphere at Titan's north pole and revealed an astounding landscape of liquid methane lakes. Scientists now conjecture that methane rains into the lakes from the methane clouds, as water does on Earth. The difference is the temperature of the rain, a frigid -180 C.

Like Io in Jupiter's magnetosphere, Titan supplies much of the mass that fills the outer magnetosphere of Saturn. As it orbits the planet at about 20 radii, its atmosphere ejects atoms of hydrogen and nitrogen, which are subsequently ionized.

Dougherty and her magnetometer team have obtained fresh information on how Titan's ionosphere interacts with Saturn's field. First they determined the moon has no magnetic field of its own; it cannot have an internal dynamo. Nevertheless, they discovered that Saturn's magnetic field piles up and drapes around the moon as it orbits the planet. The result is a tiny "induced" magnetosphere, complete with magnetotail, that is similar to those of Venus and Mars. There is no end to the marvels in Saturn's treasure chest.

Exactly How Fast Does Saturn Rotate?

It's easy to determine the rotation period of a planet like Mars that has a solid surface. Just pick a prominent feature, like a crater, and measure how long it takes to return to the same position with respect to some reference. Jupiter and the giant planets beyond it have no fixed surface features, however; only constantly changing cloud tops. And yet, the rotation period of a planet is a vital statistic, essential for understanding many of its properties. What to do?

During the Voyager flyby in 1980, radio astronomers determined a rotation period by measuring periodic variations in Saturn's kilometer-wavelength radio emission. The radiation is emitted by energetic electrons that are trapped in Saturn's dipole field. As the field rotates, the frequency of the emission varies slightly. And because the dipole field is presumably anchored deep inside Saturn, its rotation should represent some average for the planet. This method yielded a rotation period of 10 hours, 39 minutes, and 22 seconds and was accepted as a standard for twenty years.

However, radio observations from Cassini in 2003 and 2004 showed large variations of the rotation period over intervals as short as a few months. The average period was about 10 hours, 45 minutes, 45 seconds, 6 minutes longer than the old standard. Don Gurnett at the University of Iowa and his Radio Plasma Wave Science team had no explanation for this curious effect.

David Stevenson at Caltech offered his opinion that Saturn could not have slowed down by 6 minutes since the 1980s. The planet is simply too massive.

So how do we explain the difference in periods? Stevenson suggested that we might need to revise our estimates of the size of Saturn's inner core of rock and ice. But he was plainly puzzled by the results and had to await further observations.

An alternative to the radio method has always been available in principle. One could measure the rotation of the dipole magnetic field *directly*, without using the radio emissions as a proxy. This scheme works well for a planet like Jupiter, where the dipole is tilted with respect to the rotational axis, but Saturn's dipole and rotational axes differ by less than a degree, which raises the difficulty of applying the method.

Giacomo Giampieri and his colleagues at the Jet Propulsion Laboratory have finally overcome the technical obstacles. They used Cassini magnetometer data accumulated over fourteen months to extract a weak rotational signal of the field. They determined a period of 10 hours, 46 minutes, and 6 seconds that was stable to within 40 seconds for over a year. Giampieri won't go as far as saying this is the true rotation period of Saturn's dynamo field, but that it must have a close connection to it. Time will tell.

THE GIANT TWINS, URANUS AND NEPTUNE

Far out in the suburbs of the solar system, these two giant planets leisurely orbit the Sun. They each have some history. Uranus had some trouble in being accepted to the family of planets. Neptune, for its part, caused a violent controversy between French and British astronomers.

Astronomer Royal John Flamsteed was the first to see Uranus in 1690, but he failed to notice its motion and recorded it as a star. In 1776, Pierre Lemonnier observed the planet eight times in four weeks, but he too failed to see it move and overlooked its significance. Then in 1781, Sir William Herschel saw it as a disk but thought it was a comet. Charles Messier, the famous French comet hunter, knew better and declared it to be a planet. Herschel eventually agreed and named it in honor of his king, George III. The planet was also known for a time as Herschel, but the name was changed to Uranus, which is in line with the other Roman gods.

Neptune's discovery was a triumph of theory. French astronomer Alexis Bouvard noticed deviations of Uranus's motion from the orbit he had previously calculated in 1821. He proposed that an eighth planet was lurking somewhere beyond Uranus. In 1843 British astronomer John Couch Adams calculated an orbit for the postulated planet but couldn't interest anyone to look for it. Meanwhile Urbain Le Verrier independently calculated an orbit and convinced Johann Gottfried Galle at the Berlin Observatory to search. Galle found the planet in September 1846, close to the position Le Verrier had predicted. That event launched a heated debate among French and British astronomers about the true discoverer, Couch or Le Verrier. (Why not Galle?) Most impartial astronomers were willing to divide the credit equally, and that settled the matter.

Both planets are about four times the size of Earth and have about fifteen times Earth's mass. (Recall that the corresponding figures for Jupiter are eleven and three hundred, respectively.) Uranus is exceptional among the planets because its axis of rotation lies nearly in the plane of the ecliptic. So as the planet circles the Sun in eighty-four years, each polar hemisphere faces the Sun for a full forty-two years, while the other hemisphere remains dark. Neptune's axis is tipped by a more moderate 30 degrees and has four seasons similar to Earth's, but each lasts forty-two years. Both planets rotate fairly rapidly, in about seventeen hours. Both are equipped with large families of moons and rings.

Like Jupiter and Saturn, the atmospheres of the twins are composed mainly of molecular hydrogen, helium, and methane. From their oblateness, rotation, and gravity we get some clues to their interior structure. Planetary astronomers think they consist of a rocky core and a thick mantle of water, methane, and ammonia ices.

Voyager 2 was the only spacecraft to fly past these planets, so our knowledge of their magnetic fields is comparatively sparse. In 1986, Norman Ness and his colleagues at NASA's Goddard Spaceflight Center summarized their findings on Uranus and in 1989, on Neptune. Their results were unlike anything anyone could imagine.

Uranus's field can be crudely represented by a dipole, they said, whose axis is tipped from the rotational axis by 60 degrees and whose center is displaced by about 0.3 radii. Or to put it differently, quadrupolar and octupolar components are comparable in strength to the dipolar field. The solar wind distorts the field into the usual shape of a magnetosphere, with a bow shock at 24 radii upstream and a magnetopause at 18 radii (fig. 7.3 top). The field strength at 1 radius was 0.1 gauss on the night side and 1.1 gauss on the day

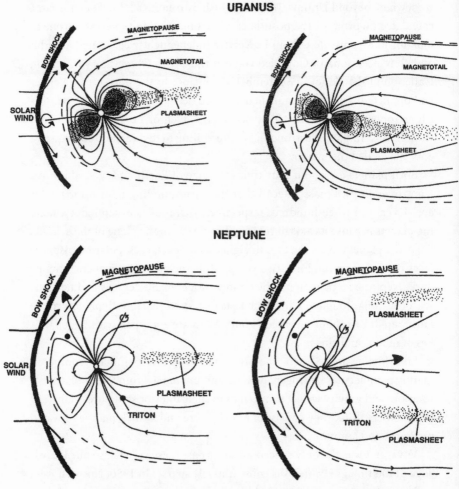

Fig. 7.3. Cartoons of the magnetospheres of Uranus and Neptune. The two panels for each planet represent the situation at opposite sides of the planet's orbit. The magnetic axis of each planet (shown with a large straight arrow) is displaced with respect to the rotation axis (shown with a curled arrow). Therefore, as the planet orbits the Sun and the rotation axis continues to point in a constant direction, the magnetic axis swivels.

side, comparable to Earth's field. Because of the large angle between magnetic and rotation axes, the magnetotail corkscrews behind the planet as it rotates.

Neptune's field is also complex, with strong quadrupole and octupole components (fig. 7.3 bottom). Its dipole axis is tilted 47 degrees from the rotation axis and displaced by 0.5 radii at the planet's center. The maximum field strength at 1 radius was measured at about 0.1 gauss, three times smaller than Earth's field. Neptune's magnetosphere has a bow shock at 35 radii upwind and a magnetopause at 26 radii.

The planets may be twins, but not identical twins. Neptune emits twice as much heat as it receives from the Sun, which indicates a source of heat in the interior. The heat drives the fastest atmospheric winds of any planet in the solar system, as high as 2,000 km/hr. Uranus seems to emit as much heat as it receives, which suggests that no interior source exists.

Theorists have worked hard to understand how the complex fields of Uranus and Neptune arise. Jupiter and Saturn are easy by comparison, because their interiors probably contain the magical substance, liquid metallic hydrogen. Convection currents in this electrically conducting fluid could make an Earth-like dynamo possible. Nobody knows for sure exactly what lies in the interiors of Uranus and Neptune, but metallic hydrogen is not likely. Some reasonable alternatives have been proposed, though, and with these in mind modelers have attempted to fashion dynamos that reproduce the observed fields.

One has to start with a model of the interior. How can one begin to know what the interior of a giant planet looks like? William Hubbard at the University of Arizona is one of a small group of scientists who have been working on this problem for the past thirty years. In 1980 he used the composition of the atmosphere, the gravitational field, the speed of rotation, and the flattened shape of each planet as constraints on its interior structure. He also included its heat losses and gains. From these parameters, he could estimate the pressures and densities along a radius. These helped to constrain the physical state of the internal gases.

He found that the interiors of all the giant icy planets fit a common three-layer pattern. Uranus and Neptune have iron-silicate cores of about 4 Earth masses; a slushy mantle of water, methane, and ammonia ices of about 10 Earth masses; and a molecular hydrogen and helium envelope of about 1 or 2 Earth masses. Jupiter and Saturn differ mainly from Uranus and Neptune in the thickness of their envelopes. The central temperatures in these planets could be high as 5,000 K.

To advance further, Hubbard needed more precise information on how well the interiors would conduct heat. The material there would be under enormous pressures, and nobody knew what their properties might be. William Nellis and his co-workers at the Lawrence Livermore National Laboratory were intrigued by the problem. They decided to measure these properties in the laboratory.

In the late 1980s, Nellis set up a shock tube, in which a high-speed jet of gas smashed into a test sample and raised its pressure and temperature instantaneously. In one test he used a mixture of ammonia and methane ices; in another a solution of liquid water and ammonia. The shock generated pressures as high 2,000 atmospheres and temperatures as high as 5,000 K, which were similar to conditions predicted for Uranus. In the few microseconds the pressures held, Nellis and crew measured the heat and electrical conductivities of the samples, which are critical quantities for models of heat flow and dynamos.

With Nellis's data in hand, Hubbard was able to refine his models of the icy giants' interiors. In 1991 he and Nellis investigated heat flow in the planets. Their models showed that *convection* of heat would not occur in most of the interiors of Uranus and Neptune because heat *conduction* was more efficient. But convection currents, on which a dynamo depends, might exist in a thin shell above a stable mantle.

Here was a vital clue for dynamo theorists. The dynamos of the giant planets might be located in a thin shell close to the surface, rather than deep in the interior. Sabine Stanley and Jeremy Bloxham, two researchers at Harvard University, picked up the idea. Professor Bloxham had been working on numerical dynamos for twenty years; Stanley, originally from the University of Toronto, was his latest graduate student.

In 2004, they reported the results of their dynamo calculations in a groundbreaking article in the journal *Nature*. They showed that a dynamo operating in a thin shell could reproduce the main features of the fields of Uranus and Neptune, including the strong quadrupole and octupole components and the lack of alignment between magnetic and rotation axes.

They investigated a variety of three-layer structures that consist of a small solid or liquid core; a stable liquid mantle; and at the top, a convecting liquid shell. As they varied the relative thickness of the shell and the electrical conductivity of the core, they could change the field from a dipolar, axial field like Earth's, to a nondipolar, asymmetric field like Uranus's. The model that best fits observations of Uranus consists of a nonconducting core one-sixth the radius in size, a nonconvecting liquid mantle one-half the radius, and a convecting liquid shell one-third of a radius.

Their results could point the way toward a comprehensive theory of planetary magnetic fields, no mean accomplishment. There is still much to be done, though. Sabine Stanley cautioned that a lot depends on how vigorous the convection and how stable the mantle are. A proper theory for these factors still lies in the future.

MAGNETIC FIELDS AND
THE BIRTH OF STARS

IN 1995, THE HUBBLE SPACE TELESCOPE captured intimate views of several stars in the throes of birth (fig. 8.1). In the top two panels, we see young stellar objects, each surrounded by a flat rotating disk and each ejecting a jet that stretches for billions of kilometers. Hundreds of such objects are forming at this moment in the Orion Nebula and others like it in the galaxy.

Astronomers are grappling with the complex physical processes that transform a diffuse cloud of gas into a normal star. Among the most challenging questions is the role of magnetic fields. Do interstellar magnetic fields help or hinder the collapse of a gas cloud? Do they influence the formation of the stellar disks? Once a protostar forms, what role does the field play? Do fields collimate the narrow jets, for example? Do they take part in generating stellar winds?

Two schools of thought have arisen over the past twenty years. One group of researchers considers magnetism to be an essential factor in star formation, second only to gravitation. The other group downplays the role of magnetic fields and places interstellar turbulence as the prime factor. Both groups have been buttressing their arguments with new observations and with complex numerical simulations. A synthesis of the two points of view may be emerging, but it is too soon to tell. In this chapter we'll compare the strong and weak points of the two scenarios.

We might keep in mind a comment made by Hendrik van de Hulst, a Dutch astrophysicist. He said, "Magnetic fields in astronomy are like sex in

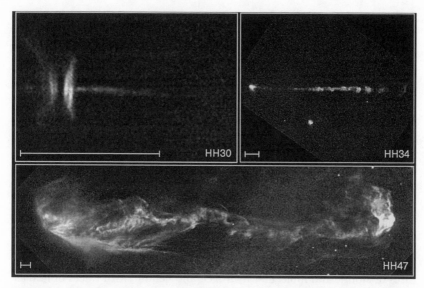

Fig. 8.1. Rotating disks and supersonic jets in stellar objects less than a million years old. These images were obtained with the Hubble Space Telescope. The bar at the bottom of each image is 1,000 AU units long. HH 30 lies in a star-forming region in Taurus, HH 34 in Orion, and HH 47 in the Gum Nebula.

psychology. Before Freud, it was ignored; then immediately afterward it was going to explain everything; but with time a more balanced view has evolved" (note 8.1).

Let's begin with some background on the nurseries where stars are born. The Orion Nebula is a good example.

ORION

One can easily see the nebula as a greenish blur in Orion's sword. It's just around the corner in our galaxy, a mere 1,500 light-years from us. (A light-year is the distance a photon travels in a year, equal to 9.46 trillion kilometers.)

When we look at the nebula with a large telescope, we can see clouds of glowing gas, studded with bright, hot stars. But this is only half the story. When we look in infrared light, we pierce the veils of dust that hide the so-called protostars that are forming as we speak. Astronomers estimate that as many as seven hundred protostars are forming in the nebula at this very moment.

The nebula is huge, about 15 light-years across, but is part of an even larger complex, the Orion Giant Molecular Cloud. This monster is at least 60 light-years in size and may contain enough molecular hydrogen to form as many as a million Suns. As seen from the Earth, it stretches twenty times the width of the full Moon.

In the beautiful images of the nebula obtained by the Hubble Telescope, we can see blobs of gas with a wide range of sizes. Astronomers have given some of them distinct names. There are individual molecular *clouds* (about 15 light-years in diameter, with particle densities of 10^3 cm^{-3}); *clumps* (about 1 light-year and 10^4 cm^{-3}); and *cores* (less than a tenth of a light-year, with densities greater than 10^4 cm^{-3}). These objects are constantly forming, merging, and dispersing. Their lifetimes depend on their sizes, with bigger objects lasting longer. A large cloud, for instance, may persist for a few million years.

We now know a lot about the gases in the Orion Nebula. They are mixtures of atomic and molecular hydrogen and helium, with traces of heavier molecules, such as carbon monoxide (CO), carbon hydride (CH), and hydroxide (OH) (note 8.2). About 1 percent of the nebula's mass consists of solid "dust" particles, about a micron in size. These motes consist of silicates, water ice, and hydrocarbons. The dust hides young stars by scattering and absorbing their light. In the process, the dust warms up and radiates in the infrared.

In so-called H II regions near a fully formed star, hydrogen is exposed to ionizing radiation. Here atoms break up into plasma of protons and electrons. Temperatures in the nebula vary from about 10 K at the borders to as much as 10,000 K near a star. The particle densities are very low, varying from one or two molecules per cubic centimeter at the fringes of a cloud, to as much as 10,000 molecules per cubic centimeter in the densest cores. (For comparison, the best vacuums we can produce in the laboratory correspond to 10 million atoms per cubic centimeter.)

Clouds in the nebula are moving randomly with speeds of less than a kilometer per second. At the low temperatures in a cloud, such motions are nevertheless supersonic and can create shocks. Astronomers often refer to such a situation as "supersonic turbulence," but at the low particle densities the behavior of the clouds is far from the turbulence observable in the laboratory. As we shall see, however, some theorists consider this type of turbulence as critical to star formation.

The nebula is permeated with a weak disordered magnetic field. Astronomers estimated the strength of galactic fields as early as 1950. They observed that starlight is very slightly polarized, presumably by needle-shaped grains of interstellar dust that align along a magnetic field. Fields of a few microgauss were common. Later, observations of radio sources through an interstellar cloud increased these estimates. Recently, millimeter wave observations of very dense cores indicate fields of tens to hundreds of microgauss.

The specific sites in the galaxy where stars form were discovered only recently. In the 1940s, Mexican astronomer Guillermo Haro and American astronomer George Herbig independently discovered bright nebulas so small as to appear star-like. (The objects in fig. 8.1 are labeled HH to indicate that Haro and Herbig discovered them.) Around the same time, Bart Bok and Edith Reilly at Harvard University discovered intensely dark globules less than a light-year in diameter, backlit by distant bright nebulas. Theorists speculated that both types of objects were possible sites of star formation.

By the 1950s, a variety of observations were suggesting that the stars in the Orion Nebula are some of the youngest in our region of the galaxy. Among these was the discovery that a group of stars was dispersing from a common point and probably originated there 2 million years ago. The so-called T Tauri stars, in the constellation Taurus, had estimated ages of only 1 million years. These were stars in the making, protostars, which were embedded in thick clouds of interstellar dust.

The breakthrough came in 1967, after sensitive infrared detectors became available. Eric Becklin and Gerry Neugebauer at Mount Wilson Observatory discovered a point-like infrared source in the Orion Nebula that was absent at visible wavelengths. Its spectrum yielded a temperature of 700 K. It was too cold to be a star, too hot to be a normal interstellar cloud. They suggested the object was a protostar surrounded by a cool shell. Douglas Kleinmann and Frank Low followed up by finding an extended infrared source that surrounds the Becklin-Neugebauer point source. Its temperature was about 70 K. They announced that theirs was the first direct observation of an interstellar cloud undergoing rapid contraction.

Since the 1970s, astronomers have depended on infrared observations to follow the various stages of star formation, from the first signs of the collapse of a clump to a core and then the further collapse of the core to a protostar. Although many of the details are still obscured by the intervening dust, astrophysicists are constructing scenarios to account for the evolution we see.

STAR FORMATION

Theorists consider the formation of stars one of the most complex and challenging problems in all of astrophysics. It involves hydrodynamics; radiative transfer in a dusty environment; turbulence; and magnetic forces. The process is fully three-dimensional and involves a wide range of time scales. The initial conditions, like the rotation of a cloud, are poorly known. Researchers also disagree on which observations are most relevant. As a result many different avenues are being explored, by comparing observations with numerical simulations. The journals are filled with different ideas.

A few general observations have served as guides for the theorist. First, stars are rarely born alone. Most stars are born as members of a binary or a small multiple system or as part of a cluster that may include dozens of stars. Second, only a small percentage of a cloud's mass ends up in protostars before the cloud disperses. In other words, star formation is a surprisingly inefficient process. Finally, the maximum lifetime of a discrete cloud is still debatable but probably is less than 10 million years.

Workers in the field generally agree that a star is born in four steps. First, an interstellar cloud must fragment somehow into dense cores in which gravity can overwhelm all other forces. Second, each core collapses, starting from the inside and progressing outward. A rapidly spinning protostar forms at the center, surrounded by a flat rotating disk. The disk must shed nearly all its angular momentum and much of its magnetic field before it can collapse into the protostar. A variety of possible mechanisms operate to generate a stellar wind that flows out of the disk and over the poles of the protostar, carrying off its excess spin. In the third phase, the protostar accretes mass rapidly from its disk. At some point, and for controversial reasons, growth stops. Finally, thermonuclear reactions ignite in the interior and the star is complete.

Gravity is the dominant force in the birth of a star but not the only one. Other factors, such as rotation, turbulence, and magnetic fields, may also be important at different stages of the birth. The cooling of the gas, assisted by the radiation from dust, is also a factor. Theorists are debating which of these secondary agents is most important, and two main schools of thought have developed. One group of researchers is convinced that magnetic fields dominate, especially in the early stages. For instance, they think the field allows the cloud to contract only very slowly, over millions of years. Cores form only after the field decouples from the gas.

The other school of thought is persuaded that magnetic fields are too weak to prevent contraction. In this view, turbulence dominates star formation by creating small density fluctuations that are able to collapse from their self-gravity. A tiny core then accretes matter from its surroundings and grows very rapidly. The whole process takes less than a few million years.

The magnetic school of thought developed a scenario of star formation that became the standard for twenty years, because it made testable predictions. In recent years the turbulence scenario has challenged the older paradigm and has gained equal stature. We'll compare how each school explains the four stages of star formation.

Contraction of a Cloud

Let's begin with a basic question: What determines whether a diffuse interstellar cloud will contract spontaneously? In 1930, Sir James Jeans, an eminent British astrophysicist, answered this question in the simplest possible situation. He imagined a cloud of a definite width, temperature, and density. He assumed that it is not rotating, has no magnetic field, and has no turbulent motions. He determined that if the mass of the cloud is too small, any tendency for it to collapse gravitationally is resisted by an increase in its gas pressure. But if its mass exceeds a certain critical size, named the Jeans mass (M_j) in his honor, gravity will overpower the gas pressure and the cloud will collapse. The greater the initial gas density and the lower the gas temperature, the smaller is the critical mass.

For example, a cloud with a density of, say, 10 hydrogen atoms/cm^3 and a temperature of 10 K must have a minimum size of about 10 light-years to collapse under its own gravity. Its critical mass will then be 150 solar masses. But this is far too large for a typical star; only a tiny minority of stars have masses as large as this. Therefore, some other factor must allow such a cloud to contract and break up into smaller masses. Or alternatively, the cloud must start out with a higher particle density.

The critical size for a cloud to contract is even larger when one includes magnetic fields. Leon Mestel recognized the problem as soon as he read the first estimates of the field strength in the galaxy, a few microgauss. Mestel was an applied mathematician who had done important work on the theory of stellar structure and evolution. In 1956 he was on a fellowship at Princeton University, working with Professor Lyman Spitzer. At the time, Spitzer was deep into a study of the physics of the interstellar medium, and between them they had an impressive expertise. They agreed to collaborate on the fragmentation of a magnetized cloud and ended by writing one of the fundamental papers on star formation.

They considered a cloud in which the magnetic pressure exceeds the thermal and turbulence pressures of the gas and derived the minimum mass it must have to enable its self-gravity to seize control. They obtained an elegant result: the critical mass (M_m) is equal to the total magnetic flux Φ (that is the number of lines of force) that crosses the midsection of the cloud, multiplied by a universal constant: $M_m = \text{Const.} \times \Phi$ (note 8.3).

For a frozen-in field of 1 microgauss and a density of 10 particles/cm^3, the minimum mass is about 440 solar masses, three times as large as the corresponding Jeans mass. They concluded, in agreement with Jeans, that unless some new factor intervened, a star as small as 1 solar mass would never be able to form in a magnetized cloud.

Mestel and Spitzer realized that one way out of their dilemma was to question their assumption of a frozen-in field. So they reexamined the microscopic processes that couple a field to a gas. They calculated that in a typical cloud, cosmic rays actually ionize only a tiny fraction of the mass. Most of the gas is electrically neutral and would be able to "slip through the lines of force" under the influence of gravity, *except* for the friction caused by the plasma (note 8.4). The neutrals would separate from the plasma and the field to which it is tied, in a process called ambipolar diffusion. In effect a cloud of *any* mass could contract slowly, until the field-free neutral gas was sufficiently dense to collapse catastrophically into a protostar. Here was a potential path to building stars smaller than the critical mass M_m.

Nevertheless, the speed at which the neutral molecules could collect by diffusion is much slower than the speed in a gravitational free fall. A cloud would contract imperceptibly over millions of years before a protostar could be formed. Ambipolar diffusion and the slow pace it imposes on cloud contraction would become a central tenet in the magnetic paradigm, and a central point of contention with the turbulence school.

Some Simple Examples

Mestel went home to Cambridge to think about the essential processes in the formation of stars. He was working in the 1960s when fast electronic computers were rare and user-unfriendly. He might not have resorted to them even if he could. He had been trained as a mathematician and was accustomed to choosing simple illustrative problems that he could solve "algebraically," that is, by hand. This style of research is productive if the researcher has sufficient physical insight. Mestel was well equipped. In 1965 he published a summary of the problems he had investigated.

In one instance, he imagined a cloud in which a magnetic field is initially uniform and straight. If the field were frozen-in, the gas would be free to contract gravitationally *along* the field lines but would be restrained from moving *across* the lines. As a result the cloud would flatten into a thin disk, in which gas pressure would balance the force of gravity.

Mestel showed that the increased gas density in the disk lowers the critical mass (M_m) for collapse (see note 8.3). He argued that the disk would therefore fragment spontaneously into pieces whose diameter is comparable to the thickness of the disk and whose mass is a few solar masses. Here was a possible solution to the problem of cloud fragmentation in a frozen-in field, but it would need to be demonstrated numerically.

Next he considered the disturbing effects of rotation. As a rotating cloud contracts, it speeds up to conserve its angular momentum, just as a skater spins faster as she pulls in her arms. Unless a cloud sheds its momentum, centrifugal forces can destroy it before it can pull itself together. The evolution of the cloud depends on the angle between the rotation axis and the straight uniform field. Mestel considered two extreme cases.

If the cloud's axis of rotation were *parallel* to the direction of the field, the cloud would wind up the straight field lines. Mestel postulated that the distant ends of the field lines were anchored in distant gas clouds. If so, the twisting motion would generate Alfvén waves that would transmit angular momentum away from the contracting cloud. In this way, the cloud could lose its angular momentum from its poles and could contract further. This simple magnetic mechanism for shedding angular momentum (magnetic braking) is one of the strongest aspects in favor of the magnetic scenario.

In the other extreme case, the cloud rotates *perpendicular* to the direction of the field. Now the cloud cannot collapse into a thin disk: centrifugal force resists the infall of gas along the field lines, and magnetic pressure resists across the field. It would seem that in such a situation, gravity is stymied; the cloud would be in equilibrium and could not contract or fragment.

Was it possible that clouds could form stars only if they happen to rotate in a favorable direction? There had to be a way out of this dilemma. Mestel realized that he had found a way, back in 1956. With ambipolar diffusion, a cloud rotating perpendicular to the magnetic field could still contract, thus solving Mestel's difficulty. Moreover, magnetic braking could dump the cloud's angular momentum.

But there is a catch. The core of a cloud must eventually lose its magnetic field as it contracts, otherwise the increasing magnetic pressure will eventually halt the contraction. So the field must be *present* for braking to occur and *absent* to allow the final collapse to a protostar. Somehow these competing processes must occur in a well-coordinated sequence. Theorists would struggle for decades to reconcile these demands in a coherent theory.

Collapse of a Core

Mestel laid out some of the essential issues in the contraction and collapse of a magnetized cloud, but in an essentially heuristic fashion. By the 1970s many other researchers were using numerical simulations to explore these issues more quantitatively. Takenori Nakano, a physicist at Kyoto University, was the first to incorporate ambipolar diffusion in a numerical model of a contracting magnetized cloud.

He started his model with a flattened cloud of 50 solar masses in a uniform, straight field of 4 microgauss. He postulated that the diffusion of neutral molecules relative to the field would be so slow at first that the cloud would be almost static. That allowed him to think of the cloud as nearly in mechanical equilibrium at every moment.

Then he calculated the rate of drift of the neutrals across the field lines. In his model the high-density center contracts faster than the outer parts. As the gas density increases, the plasma decays, because ions and electrons recombine faster than cosmic rays can separate them. Therefore, ambipolar diffusion accelerates and the center begins to collapse in a free fall. After a mere 300,000 years, the center contains 3 solar masses and a field of 10 milligauss. Nakano had to terminate his calculation at this point, but he argued that a realistic star would be born following the collapse.

Nakano's result was an important step in developing the magnetic scenario for star formation. But notice that he started out with a small cloud on the verge of collapse. He bypassed the earliest stages of contraction and ignored the possibility that the cloud might fragment into many cores. In effect he followed the late evolution of one core. Many researchers would repeat his calculations with more refinements but without tackling the problem of fragmentation.

Eugene Scott and David Black, two physicists in California, extended Nakano's results by following the evolution of a cloud from the very beginning. Their simulations of 1981 showed that the outer parts of a cloud would definitely contract more slowly than the center, whether or not the field was frozen-in. If the cloud were weakly ionized, so that ambipolar diffusion prevailed, the field would remain in the outer parts as the neutral gas drifted to the center. In effect the field would be redistributed, leaving the core almost field-free. The cloud would then separate into a rapidly contracting core and a magnetically supported envelope.

In 1987, Frank Shu, Susana Lizano, and Fred Adams advanced a cluster of ideas that spurred further development of the magnetic scenario. Shu was a professor at UC Berkeley and is now at UC San Diego. He is one of the most vigorous proponents of the role of magnetic fields in star formation. With his many graduate students, he has developed the theory for the past thirty years. Lizano's home base was the National Autonomous University of Mexico. She is now a professor of astrophysics there. Adams is a professor at Michigan State University.

Shu had previously calculated how a core would collapse into an extremely dense object, a "singular isothermal sphere" that is supported by thermal pressure alone. Gas would flow into the singularity "from the inside out." That

meant that the infall velocity would be large near the singularity and decrease to nearly zero in the surrounding envelope. This early model had a
huge impact on the subject during the 1980s and was confirmed by observations of cores in the 1990s.

In their 1987 presentation, Shu and friends proposed that a clump in a
cloud could collapse in one of two ways, depending on whether its mass is
larger or smaller than the magnetic critical mass (M_m) appropriate to its total
magnetic flux Φ. If its mass exceeds M_m, it is "supercritical" and the field cannot prevent the clump from collapsing rapidly, *even if it is frozen in*. In this
case the clump would collapse, as Mestel predicted and as Black and Scott
had demonstrated: preferentially along the field lines to form a disk. Supposedly, the disk would fragment into cores of a size comparable with the thickness of the disk, and with a mass-to-flux ratio that is also supercritical. Such
cores, they speculated, could develop into a few very massive stars, such as
the cluster of O and B stars in Orion.

If, however, a clump is "subcritical," cores could still form by ambipolar diffusion. The process would proceed slowly, over 10 million years or so. If the
cloud contained many Jeans masses (M_j), it might fragment into many cores
of approximately 1 solar mass each. The cores would ultimately form a cluster of stars that are loosely bound by gravity, like the T Tauri stars seen in the
constellation Taurus. These are among the youngest stars known; they've
ignited their thermonuclear furnaces only recently.

All this was inspired conjecture so far. Shu and colleagues never actually
demonstrated how a subcritical clump would fragment into cores, but later
they did carry out extensive three-dimensional calculations of the evolution
of a single core in a subcritical cloud. They ignored rotation but did include
turbulent pressure and, of course, ambipolar diffusion from a surrounding
envelope. Their results showed in detail how the density would slowly build
up in a central molecular condensation in which the field would be excluded.
Then a gravitational instability would cause the core to collapse rapidly.

More Realistic Simulations

By the mid-1990s, the contraction of a cloud by ambipolar diffusion was fairly
well understood. Researchers were ready to tackle the effects of rotation and
magnetic braking. All clouds rotate slowly, and as they contract, they spin

up. They must shed their angular momentum to avoid being torn apart by centrifugal forces. How would the magnetic field assist in this process?

Telemachos Mouschovias and his student Shantanu Basu were among those exploring these issues. Mouschovias had been a charter member of the magnetic school of star formation, beginning in the 1970s. He is an extremely energetic and motivated researcher who has always had access to powerful computers at the University of Illinois. He has been able to contribute many insights over the years.

Their model of 1994 consists of a rotating disk-like cloud surrounded by a static gaseous envelope. At the beginning of their simulation, a uniform magnetic field threads through both the cloud and the envelope. Then ambipolar diffusion forms a dense supercritical core, while the envelope remains supported by the field. As the core rotates, torsional Alfvén waves carry away its angular momentum to the envelope. The process is so efficient that centrifugal forces are minimized and never build up sufficiently to interrupt the collapse.

The core's angular rotation speed drops dramatically at first, then levels off, and finally rises precipitously again as the core begins to collapse in a free fall. At the end of the calculation, after an elapsed time of 10 million years, the density of the core has risen from 10^3 to 10^9 molecules/cm^3 and it is rotating ten times faster than the whole cloud was initially.

In short, the core still retains a dangerous amount of angular momentum. Basu and Mouschovias suggested that the core would break up into a binary, converting its rotational momentum to the orbital momentum of two stars. As we mentioned earlier, more than half of the stars in the galaxy are members of a binary or cluster, so that the breakup of a core seems very plausible.

The Fragmentation Problem

The collapse of a core is a favorite topic among workers in the field and has spawned a huge number of examples in the journals. Numerical simulations have been carried out to explore the effect of dust, radiation, hydromagnetic waves, rotation, and turbulence.

On the other hand, few attempts have been made to simulate the fragmentation of a magnetized cloud from its initial state. Most members of the magnetic school of thought have been content to announce that the cloud, or parts of it, have become supercritical and therefore *must* break up. But this doesn't answer an important question: "How many stars of a chosen mass form in a cloud?" Observations show, for instance, that stars of 1 solar mass are about 170 times more plentiful than those of 10 solar masses. One would like to be able to predict the distribution of masses from first principles.

Alan Boss at Yale University has studied the problem for over a decade. He's carried out a series of numerical simulations of the fragmentation of a rotating cloud in which ambipolar diffusion prevails. In 2000 he demonstrated that the tension of the magnetic field acts to prevent the formation of a single dense center in a core and leads to the formation of two, four, or more fragments. But he hasn't been able to generate a distribution of masses that can be compared with observations.

Ellen Zweibel and her colleague Rémy Indebetouw at the University of Colorado have also made some progress on the problem. In 2000, they modeled a cloud as a flat disk of weakly ionized gas that is threaded with a uniform magnetic field perpendicular to the plane of the disk. Then they imposed tiny fluctuations of density (seeds) of different lengths and allowed them to evolve under the influence of gravity, magnetic tension, and ambipolar diffusion.

Seeds with particular lengths increased their density faster than others. In different trials, the disk broke up into a few elongated fragments that collapsed, lost magnetic flux through ambipolar diffusion, and began to orbit each other.

At the end of their simulation, the fragments had masses of one to ten Suns, depending on the initial gas temperature. A fragment grew in density at a rate that is intermediate between slow ambipolar diffusion and fast free fall. As it collapsed, it developed a pair of counterrotating vortices, in which the gas is streaming vigorously along the lines of force. They make a pretty picture.

These results were encouraging, but unfortunately other researchers concluded that a magnetic field could actually *inhibit* the fragmentation of a cloud. Eric Price and Matthew Bates at the University of Exeter recently completed simulations that bear on the question. They showed in 2007 that if the rotation axis of the core is parallel to the field, magnetic *pressure* restrains fragmentation. If the rotation axis is perpendicular to the field, however, magnetic *tension* inhibits fragmentation. But if a core is perturbed sufficiently, a binary will form. Its separation depends on the mass-to-flux ratio, M/Φ. Ratios as large as 10 or 20 (corresponding to relatively weak fields) produce wide separations; small ratios produce small separations. In all cases a spiraling filament of gas joins the two fragments.

So it would seem that the fragmentation of a magnetized cloud is still a matter of some debate.

THE OTHER SIDE OF THE FENCE

It's time to turn to the turbulence school of star formation. Richard Larson at Yale University is probably the founding member of the group. In 1981 he proposed that supersonic turbulence in a cloud would produce high-density cores that would grow into protostars. In 2003, he reviewed the status of star formation theory from this standpoint and also discussed the weak points of the magnetic scenario.

He first pointed to evidence that magnetic fields do not stabilize clouds for millions of years, as the magnetic school assumes. Clouds, he wrote, are observed as filamentary, highly irregular in shape, and turbulent. These properties suggest that a cloud is changing rapidly and, as a result, that magnetic fields are weak and disordered. Second, at the time he wrote, observations of magnetic field strength and uniformity were scarce and presented a mixed picture. So he claimed that the observations don't support the assumptions of the magnetic school. (We'll return to this point later.)

Larson's strongest argument against the magnetic control of core formation is the lifetime of clouds. Stars and clusters older than 10 million years have no associated clouds. That argues for a cloud lifetime of less than 10 million years. The turbulence scenario builds cores in a few hundred thousand years. In contrast, models employing ambipolar diffusion require at least a few million years to form a core. Larson grants that magnetic fields are present and may play a crucial role in magnetic braking but asserts that they cannot control the formation of cores.

How, then, would the turbulence school explain the production of cores in a cloud? Two possible routes have been explored with numerical simulations. First, and very simply, small random clumps in a cloud might collapse from their own self-gravity and continue to pull in more mass from their surroundings. As usual, rotation, turbulence, and magnetic fields tend to resist gravitational collapse, but once a clump forms, the final collapse could occur very rapidly, say, within a few hundred thousand years.

In this picture, tiny seeds grow into mighty stars; the key process is accretion rather than fragmentation of large clumps. The seeds have a lot of growing to do, because some simulations produce cores as small as 0.01 solar masses.

Alternatively, turbulence in a cloud would produce shocks, in which the gas density would be increased momentarily ten- or a hundredfold. That could form a seed from which a core might grow. If a seed were fortuitously as large as a Jeans mass, in which gravity and gas pressure are roughly in balance, the seed might survive to become a protostar.

I've selected a couple of examples to illustrate the work of the turbulence school. We'll start with a team at Harvard University, led by Paolo Padoan. In 2000 they calculated the formation of cores in a turbulent cloud. Their results are interesting because they were able to predict how many cores are formed with a chosen mass, the so-called initial stellar mass function. This is an important test for any theory because it can easily be checked against observations. As we have seen, the magnetic school has so far encountered difficulties in predicting the mass function.

In previous work, Padoan and friends demonstrated that supersonic turbulence in a magnetic field creates a complex system of shocks that fragment the gas into high-density sheets, filaments, and cores. In this simulation, they added gravity to the problem. Their model cloud had a diameter of 45 light-years, an average magnetic field of 4 microgauss, a particle density of 900/cm^3, a temperature of 10 K, and a random turbulent velocity of 2 km/s, or ten times the speed of sound.

As a general rule, turbulence in a cloud decays in about the time a shock wave takes to cross the cloud. In this case this "dynamic time" is 5 million years. That is comparable to the lifetime observed for typical clouds, which may be a happy coincidence.

Padoan and his colleagues turned on the random turbulence in the cloud, and after 5 million years the cloud is filled with cores, with masses between 0.5 and 200 solar masses. Then they selected the supercritical cores to construct their mass function, because only these would collapse further into protostars. As expected, less-massive cores were more frequent. Their predicted mass function compares very well with the function observed in stellar clusters. This was a major success.

There was more good news. When they compared the predicted radii, rotation speeds, and densities of their cores with observations of 150 cores, they found the variations with mass were surprisingly good. In addition, only 2 percent of the initial mass of the cloud ended in cores, as observation suggests.

These researchers concluded that there is "no need" to invoke ambipolar diffusion in a subcritical cloud to produce cores of low mass, as the magnetic school has done. Turbulence, they claim, could do it all. They make a good argument.

In figure 8.2 we see the results of a similar exercise in 2005, by Anna-Katharina Jappsen and friends at the Astrophysical Institute of Potsdam. They included a more accurate treatment of the changes in the thermal state of the gas as it is compressed. The black dots are the sites of dense cores. They cluster in the sheets of high density produced by the turbulent shocks. These authors concluded that the stellar mass function depends mainly on the thermodynamic state of the gas, which is controlled by heating and cooling processes.

More than half of all stars are born as members of a binary. Can the turbulence scenario demonstrate how this occurs? Richard Kline and his colleagues at UC Berkeley have tried. They have developed three-dimensional hydrodynamic codes that can handle, for the first time, the nonlinear effects of the density and velocity variations within a cloud core. They include rotation in their calculations, but they ignore magnetic fields.

In one simulation, they began with a turbulent isothermal core that is just on the verge of gravitational collapse. They specified the strength of turbulence with the Mach number (M), the ratio of the "average" turbulent velocity to the local speed of sound. The gas density at the edge of the core determines the mass of the core. In one example the core mass was about 3 solar masses.

They learned that "subsonic" turbulence (M less than 1) produces a single fragment, with half the mass of the core. Here was a possible way to produce brown dwarfs, which are stars of less than a solar mass. With "transonic" turbulence, at Mach 1.7, a binary of two protostars appears, containing about 1 solar mass. Ninety percent of the original mass of the core ends up in a dense

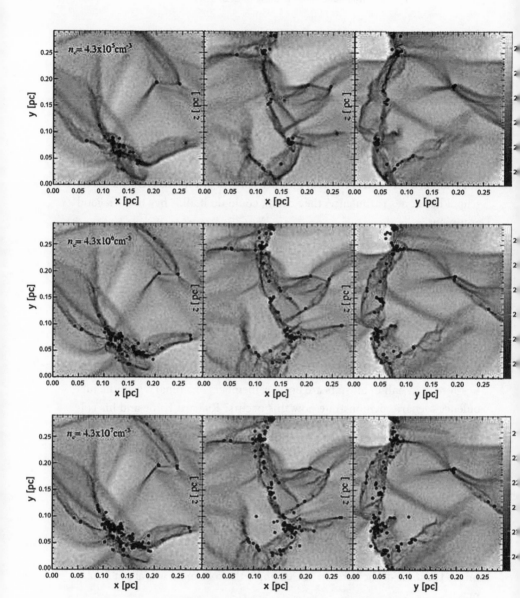

Fig. 8.2. Dense cores are formed at the intersections of hydrodynamic shocks in these simulations. The horizontal and vertical scales are measured in parsecs (pc). A parsec equals 3.26 light-years. Projections in the xy, xz, and yz planes are shown for three densities.

filament that connects the stars. Finally, they turned up the turbulence knob to full strength, Mach 3. Another binary developed, with a total mass of 20 percent of the original 8 solar masses. At present they haven't been able to run their programs long enough to see what happens to the massive filament. But their work shows promise.

WHO HAS THE LAST WORD?

We've contrasted the approaches to star formation that two schools of thought have taken. One group sees the magnetic field as both an impediment to forming dense prestellar cores in a cloud and an aid in shedding a core's angular momentum. In this scheme, cores form *slowly* by ambipolar diffusion in subcritical clouds. The other group sees shocks in supersonic turbulence as the principal agent in forming cores. In this scenario, cores form *rapidly* in a free fall. Each approach has its successes and weak points. Can new observations decide which scenario is realized in nature?

Richard Crutcher at the University of Illinois has tried. Over the past decade he has obtained observations of the magnetic field in cloud cores to decide between these two extreme scenarios. Crutcher is a radio astronomer who observed regularly with the Berkeley-Illinois Millimeter Wave Array (note 8.5). It's an interferometer at Hat Creek, California, that offers arcsecond angular resolution and the ability to detect weak magnetic fields. Crutcher derives the magnetic field direction and strength from the polarized emission of carbon monoxide (CO) and cyanogen (CN) molecules, and the polarized emission of warm dust.

In 2004 he reported his observations of two neighboring cores with a total mass of about a hundred Suns. His key result was that the magnetic field is straight and parallel in the cores and has the huge strength of 1 milligauss, a thousand times that of the surrounding cloud. So, he concluded, turbulence must be too weak to distort the field lines. In this case, magnetic forces would control whether the core would fragment any further to form protostars. However, contrary to the predictions of the magnetic model, the shorter axis of the core didn't line up parallel to the field. That could mean that turbulence still has some influence.

The latest word on the subject comes from Derek Ward-Thomson at Cardiff University and his international team of experts. In 2006 they surveyed all the relevant observations they could find and offered some interesting conclusions.

One way to distinguish between the competing scenarios is to look at the predicted formation time of a core. The magnetic scenario predicts that a core forms in a subcritical cloud within a few million years, while the turbulent scheme predicts that one will form in a few hundred thousand years. So the team examined the lifetimes of cores and their dependence on the initial density of the cloud. They decided that the observed lifetimes lie *between* the predictions of both scenarios. Life is never easy.

Since this test was not conclusive, the team turned to observations of the magnetic field.

Only five prestellar cores have been studied in enough detail to bear on the issue. Once again the results are mixed. In three cores Crutcher studied, the fields are straight, strong, and uniform, as the magnet school assumes, but the core is not flattened along the lines of force.

Another possible indicator of the role of magnetic fields is the ratio of mass to flux (M/Φ) in a core. The magnetic school predicts that cores should be mildly supercritical (M/Φ slightly larger than 1); the turbulence school predicts much larger ratios. All five well-studied cores are on the borderline of being critical, with M/Φ about equal to 1. So this test supports the magnetic school.

The team also reviewed observations of the stellar mass function; that is, the number of protostars of a chosen mass that form in a cloud. They concluded with some caveats that the observed function does not change among clouds of different total masses, which scores a point for the turbulence school.

Altogether, the weight of evidence suggests that turbulence and magnetic fields contribute about equally in the formation of prestellar cores and their subsequent evolution. Neither paradigm can explain everything that is seen; neither is excluded by the data. But the critical observations that bear on the issue are still surprisingly sparse.

DISKS, WINDS, AND JETS

We've discussed how two competing theories explain how dense cores might form in a molecular cloud and how they collapse rapidly into a protostar. What does a protostar actually look like? Let's look again at figure 8.1.

The ball of gas that will become an ordinary star nestles in a caul of warm dust and gas and is invisible in these images. The most prominent features are the luminous jets that escape from the poles of the star and the flat accretion disk that lies in the equatorial plane. Gas from the surrounding envelope flows into the disk at its edge. Part of the gas reaches the star, while some escapes as a jet from the polar regions or as a wind from the disk itself. These flows carry off most of the angular momentum with which the gas arrived, enabling a portion of the gas to attach to the star.

Notice the scale of the images: each bar in the corner of an image represents 1,000 AU, about equal to the entire heliosphere. The core from which each protostar formed was gigantic in comparison, perhaps 3,000 times larger. The Sun-like star that will evolve from the protostar will have a diameter of only about 0.01 AU.

Much of what we've learned about protostars has been derived from studying the T Tauri stars, a cluster 460 light-years from Earth in the constellation Taurus (note 8.6). High-resolution imaging and spectroscopy, particularly from the Hubble Space Telescope and lately from the Spitzer Space Telescope, have given us a fairly detailed picture of the structure and motions within such objects. They are very young, less than a million years old, and contain only a few solar masses. At the center of each one lies a dim red star whose thermonuclear reactions have just recently fired up.

Their disks of gas and dust have a wide range of masses, from 0.001 to 1.0 solar masses. Disks don't rotate as a solid body would. Following Kepler's second law, the outer parts rotate more slowly than the inner parts. Mass is flowing into the disk from the surrounding envelope at a fraction of a kilometer per second and onto the star at 100 to 300 km/s. Some of this mass is redirected into polar molecular flows, which have speeds of 10 to 50 km/s.

The jets are really spectacular. They are composed of ionized hydrogen streaming at supersonic speeds of 150 to 400 km/s, within a narrow cone of about 3 degrees. The jet in DG Tau B extends from a distance of 30 to 1,000 AU from the star and remains collimated all the way. Jets carry off mass at tremendous rates, ten thousand to a million times the rate of the solar wind. Moreover, they tend to sweep up the slower-moving polar winds that surround them.

The T Tauri stars also have strong magnetic fields. In 2007, Christopher Johns-Krull at Rice University observed equatorial magnetic fields of about

1 kilogauss in a sample of fourteen stars. Other observers have detected fields of around 2 kilogauss. The shape of the field is controversial. Some researchers postulate a dipole stellar field, others picture only a disk field, still others a combination of the two. As we shall see, these strong fields play an important part in dynamical theories of protostars.

Theorists are trying to understand how the complicated flows into and out of a protostar originate. The most striking feature of the flows is that they are cold. Back in chapter 5 we discussed the hot solar wind that expands from the pressure of a million-degree corona. T Tauri stars also have coronas that emit x-rays, but their winds consist of cold molecular hydrogen, not hot plasma. So how are they driven? Their supersonic jets are warm enough to be ionized (at perhaps a few tens of thousands of kelvin), but they too have insufficient thermal energy to escape the protostar. What is the source of their energy?

These questions are tied up with two other major issues. The first concerns the so-called angular momentum problem. If a T Tauri star were to acquire all the angular momentum of its parent core, it would spin fast enough to disintegrate. Yet the T Tauri stars rotate relatively slowly, with periods of a few days. Somehow the disks and jets carry off the unwanted momentum. How is this done? Similarly, if the star were to retain all the magnetic flux of its core, it would end up with a horrendously large surface field, far in excess of the observed field. How does it manage that?

Roger Blandford and Donald Payne, two astrophysicists at Caltech, offered answers to these questions in 1982. They pictured a differentially rotating disk that is threaded with a weak magnetic field. The partially ionized gas flowing through the disk carries the field toward the star. Assuming that the gas is cold and has zero resistivity, the authors showed how the centrifugal force would fling some of the gas away from the disk along the field lines, like a bead on a whirling wire. For this to occur, the field lines must tilt away from the surface of the disk by an angle of 60 degrees or less.

The differential rotation of the disk twists the field lines so that, at large distances from the disk, the field acquires a strong toroidal component. In effect, the disk wraps its inner field around the axis of rotation and forms two helices. They collimate the flow into two antiparallel polar jets. Most of the energy of the jet is concentrated in its core, while most of the angular momentum and magnetic flux resides in the jet's walls. Blandford and Payne suggested this model to account for accretion disks and jets around black holes, but it served as well to describe young stellar objects.

This basic model has been modified and extended by a large number of researchers. For example, Frank Shu and colleagues argued in 1994 that most of the magnetic field in the disk would be crowded inward to the "co-rotation radius," where the gas is forced to rotate with the star's dipole field (fig. 8.3). At this radius the field lines open to space and guide a centrifugally driven wind. The remainder of the inflow can funnel into the stellar field and onto the star.

Shu and others like him search for steady-state models, in which the flows would remain stable indefinitely. Ralph Pudritz at McMaster University and Arieh Königl at the University of Chicago, among others, have devoted much effort to explain the stable interaction of a disk, a star, and a jet. All of these researchers use complex numerical simulations to explore such factors as the conductivity and turbulence of the gas and the inertial effects of mass on the field lines.

Nevertheless, Lee Hartmann at the University of Michigan, among others, has questioned whether such stable flows are realistic. He pointed to FU Orionis, a young stellar object that experiences dramatic variations of mass accretion, accompanied by huge increases of brightness. Also, discrete blobs are seen moving out along the jets of some young objects (see HH 34 in fig. 8.1), which strongly suggests some kind of erratic behavior. Another school of thought has developed, therefore, that sees star formation as occurring in successive bursts of accretion.

In some of these models the reconnection of field lines enables a magnetic bubble to detach completely from the disk and carry off angular momentum, mass, and energy. Later the field lines around the disk close by reconnection, and the buildup of a bubble begins anew. The whole sequence may require several years to complete. Anthony Goodson at the University of Washington and his international team have recently demonstrated this episodic scenario with numerical simulations.

Theorists still don't understand in any detail how the jets are propelled or how they remain tightly collimated over distances of several light-years. Numerical simulations by Königl, Pudritz, and others show how the jet might form near the star as the disk wraps its field around the rotation axis. The

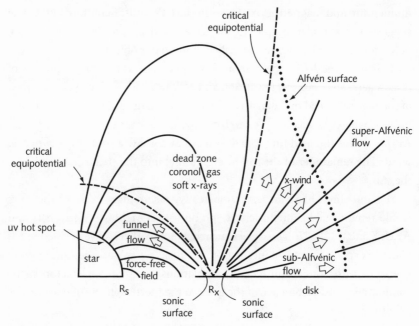

Fig. 8.3. A cartoon of the flows near a classical T Tauri star. At a critical radius where the stellar dipole field meets the inner disk, magnetic field lines open to space and release a centrifugally driven wind. Some gas crosses the radius and funnels onto the star. This model doesn't address the origin of a jet.

energy to launch the jet is essentially gravitational. The star's gravity acceler-ates the gas in the disk, and some of this energy is transferred to the jet by means of the magnetic field. But the persistence of the jet over long distances remains a serious problem.

In 2005, Franco Bacciotti and an international team obtained observa-tions with the Hubble Space Telescope that may help to test some of the mod-els, at least for the inner parts of a jet. They observed the inner 200 AU of the jet of DT Tau, a classic T Tauri star. From a series of spectra, they were able to measure the profiles of the velocity and density structure across the width of the jet.

The width at the base of the jet was a few AU, and collimation was achieved within 10 to 20 AU. The jet's velocity is highest in its core (up to 400 km/s) and decreases to about 70 km/s at the outer edge. These properties, as well as the estimated mass flux, are consistent with predictions of some magnetic-centrifugal models.

The really interesting findings concern the rotation of the jet. According to the models, jets carry off a substantial fraction of the disk's angular momentum, so that some signs of rotation should be present. Bacciotti and friends found them in DG Tauri and RG Aurigae. The outer edges of the jets rotate at speeds of 10 to 25 km/s, in accord with the models. That would account for about 60 percent of the disk's angular momentum.

Bacciotti and friends were also able to estimate the radius in the disk at which the observed part of the rotating wind is launched. This turned out to be between 0.5 and 2 AU from the star, too wide a range to agree with Shu's 1994 model.

Models like those of Königl and Pudritz (2000) explain the collimation of a jet by means of the "hoop stress" of the jet's helical field. In effect the toroidal component of the field wants to contract toward the axis of the jet and is balanced by the jet's internal pressure. Bacciotti's team was able to confirm this scenario by analyzing the profile of the velocity across the width of the jet. This was good news for the modelers.

The supersonic jet, however, is still at the frontier of this research. One wonders, for example, why a long jet doesn't become unstable, develop wiggles, and collapse. It is a hot topic at conferences these days. But at least we have a beginning of some understanding.

ABNORMAL STARS

BACK AROUND 1896, ANNIE JUMP CANNON was one of a group of women working for Dr. Edward Pickering, director of the Harvard College Observatory. He had them sorting stars according to their spectral characteristics, with the aim of compiling a comprehensive catalog. The catalog would be dedicated to the memory of Henry Draper, a prominent New York doctor and astronomer.

Miss Cannon was the acknowledged expert of the group. Her fabulous eye and quick mind enabled her to distinguish the smallest gradations among the spectra. She conceived a classification scheme that has lasted to this day, with bright blue-white stars (labeled O, or B) at one end and dim red stars (M) at the other. The Sun is a yellow G star (note 9.1). She's credited with examining and classifying most of the 400,000 stars in the now-famous Henry Draper catalog.

Most "normal" stars fit smoothly into her classification scheme, but some were "peculiar." They had a few spectral lines that were extraordinarily strong or weak, compared to the normal stars of the same spectral type. For example, some white stars labeled type A in her scheme had lines of iron or silicon hundreds of times stronger than normal. In one star the lines of strontium were a thousand times stronger. She designated such stars as "Ap," the little "p" standing for "peculiar." They would remain something of a mystery for another fifty years.

Horace Babcock, of the Mount Wilson Observatory, helped to dispel some of their mystery. In the 1950s he was busy exploring the magnetic properties of active regions on the Sun. (Recall our discussion of his work in chapter 3.) Naturally, he was curious whether other stars might also possess detectable fields. So for over a decade, he spent his spare nights observing stars of various spectral types at the 100-inch and 200-inch telescopes.

Babcock used an optical analyzer and a spectrograph to search for the splitting and polarization of spectral lines that are sensitive to the Zeeman effect. (Recall that when an atom is subjected to a magnetic field, the lines it emits split into distinctive patterns, whose width in wavelength indicates the strength of the field.) He chose stars with narrow spectral lines, so that the small Zeeman effect he was searching for might be detectable. Normally, a star's rotation broadens the spectral lines it emits by an amount that could easily swamp the effect (note 9.2). What Babcock was doing was selecting stars whose axes of rotation were nearly aligned with his line of sight.

By 1958 he had collected eighty-nine stars that showed a magnetic signal. All had magnetic fields, averaged over their surfaces, of about a kilogauss. One of them, known thereafter as Babcock's Star, had a whopping field of 34.4 kilogauss. Compare that to the Sun, whose globally averaged field strength is less than a gauss.

The big surprise was yet to come. Babcock discovered that the great majority of his magnetic stars were peculiar A or B stars. Here was a definite connection between strong magnetic fields and spectral peculiarities. But Babcock also found a few Ap stars without detectable magnetic fields and a variety of stars with strong fields and no spectral anomalies. So the connection is complicated. Nevertheless, Babcock surmised that most stars probably have magnetic fields, although only those with strong coherent fields might be detectable.

Among Babcock's magnetic stars were some whose field strength varied periodically in four to nine days. Some of these even reversed polarity during their cycle. Two explanations were proposed for this behavior: a rapid oscillation of the whole field, similar to the twenty-two-year variation of the Sun; or a rigidly rotating dipole, with the magnetic and rotational axes separated by a small angle. The latter explanation fit the data better and eventually was accepted as the "oblique rotator" model.

By the 1960s, a theory for the formation of spectral lines in a stellar atmosphere was in hand. Researchers would apply it to show that the anomalously strong or weak lines in Ap stars represented real differences in the concentration of chemical elements in their atmospheres. In some Ap stars, iron and chromium were a thousand times more abundant than in the Sun, which is the standard "normal" star. In a famous europium star, the element was a million times more abundant.

So several questions arose. What is the cause of these huge abundance differences? Are they related to the presence of a strong magnetic field? And how do these stars develop such strong global fields?

Georges Michaud, a Canadian graduate student at Caltech, offered a reasonable answer to the first question in 1970. He assumed that A stars have quiet, stable atmospheres because, he thought, they are slow rotators and have shallow convection zones. He calculated that heavy ions like iron would sink to the bottom of such a stable atmosphere unless a force larger than gravity drove them aloft. Then he showed how radiation pressure in these hot stars could provide the necessary force to lift certain ions. Typically these ions have many spectral lines at wavelengths where most of the stellar radiation is concentrated. The ions would absorb the momentum of photons and would diffuse toward the top of the atmosphere, where their spectral lines are observed. An apparent overabundance of these ions would result from this process. Ions that don't absorb sufficient radiation selectively would drift toward the bottom of the atmosphere, thereby producing an apparent underabundance near the top.

Michaud's theory of diffusion was only one of several ideas floated in the 1970s to explain the extraordinary abundance observations. Among these were: element synthesis deep in the interiors of Ap stars; contamination of the surface by nearby supernova explosions; and selective accretion from the interstellar gases. Only the accretion scenario took account of the strong magnetic fields on Ap stars. Michaud himself considered their magnetic fields to be relatively unimportant in producing the abundance anomalies. That conclusion would be sorely tested later on.

In the following twenty years, the magnetic fields of dozens of Ap stars were measured, with methods similar to those of Babcock. John Landstreet at the University of Western Ontario was a pioneer in this effort. Collaborating with Ermanno Borra, he completed a survey in 1980 of the forty brightest Ap stars in the northern hemisphere. They recorded the time variation of the circularly polarized light in the wings of a strong hydrogen line, H beta.

For each star, they derived an oblique rotator model of a dipole field and determined both the polar field strength and the angle between the magnetic and rotation axes. Field strengths as high as 5 kilogauss (0.5 tesla) were observed in slow rotators (periods of three to ten days), while the faster rotators (with periods around one day) were limited to a few hundred gauss. A simple dipole, with its center displaced from the center of the star, could represent the data quite well.

As observations improved, however, more complex models were invoked to represent the data. In 1990 Landstreet collaborated with Gautier Mathys (European Southern Observatory, Chile) in analyzing the high-quality observations of twenty-four Ap stars that Mathys had obtained over the previous decade. They showed that the fields of these stars really were combinations of dipole, quadrupole, and octupole components.

STARSPOTS

In a parallel development, researchers discovered stars that flared or showed violent chromospheric activity similar to that of the Sun. The star BY Draconis was a prime example. It is a binary consisting of a dwarf K and a dwarf M star; two red stars that orbit in about six days. The M star erupts occasionally in a violent flare, brightening by a factor of 2 to 4. In addition, its brightness varies semiperiodically, with slow changes in amplitude, phase, and mean light level. These slow variations were attributed to dark "starspots" that rotate on and off its visible disk.

RS Canes Venatici is another famous spotted star. It too is a binary, consisting of a K1 subgiant and a G5 subgiant. The K star has varied periodically in brightness by as much as 40 percent, for as long as a hundred periods. That implies that the area covered by a dark spot must have exceeded 15 percent of the total global area. Now that is indeed a spot to be reckoned with! For comparison, the largest sunspot ever seen measured 0.6 percent of the solar disk.

In the 1970s dozens of spotted stars were discovered from their periodic variations of brightness. The amplitude of the variation yielded the fractional area of the spot, and when light curves in several colors were compiled the temperature of a spot could also be deduced. But the light curves revealed nothing about the latitude of the spot and very little about its shape.

In 1980 Steven Vogt and Donald Penrod, both at the Lick Observatory, made an important advance. They invented a technique they called Doppler imaging that allowed them to determine the latitudes of starspots, among other interesting properties. The method applies to rapidly rotating, spotted stars such as RS Can Ven.

Here's how the method works for a particular spectral line. Each area on the star emits a narrow absorption line, which is Doppler shifted in wavelength as the area rotates around the star's axis. As the area rotates toward the observer, its spectral line shifts toward the blue; as it rotates away from the observer, the line shifts toward the red. All areas on the stellar disk emit simultaneously in this way. Therefore, their narrow shifted lines add up to form a much broader line that the observer sees (fig. 9.1).

This broadened line doesn't change if the star lacks a starspot. But if a dark spot appears on the visible stellar disk, a little bump appears on the profile of the spectral line. The bump moves in wavelength across the line profile as the spot rotates across the disk. If one observes the bump for several rotations, one can determine its linear speed of rotation and therefore the latitude of the starspot. A single spectrum is sufficient to determine the starspot's longitude.

Vogt and Penrod inaugurated their method on RS Can Ven (otherwise known as HR 1099), because it has several of the necessary prerequisites. It's the brightest of its class of stars, it rotates rapidly, and its rotation axis is tipped from the line of sight by a convenient 33 degrees. They discovered two dark spots, one at low latitude and a huge one at the rotational pole of the star. The polar spot had a long extension to lower latitudes. They estimated that it covered between 8 and 10 percent of the surface, which is huge by solar standards.

Of course, the Sun has no polar spots, so this discovery was quite surprising. At first other researchers refused to believe it and searched for alternative explanations, such as contamination of the observations. Theorists were intrigued, though, because it suggested that a novel form of stellar dynamo might be in operation.

As technology improved, observers could resolve several starspots on a star. In 1993 Artie Hatzes at the MacDonald Observatory of the University of Texas was able to map the size, shape, and location of five spots on the RS Can Ven star, Sigma Geminorum. They occupied an active band at 55 degrees

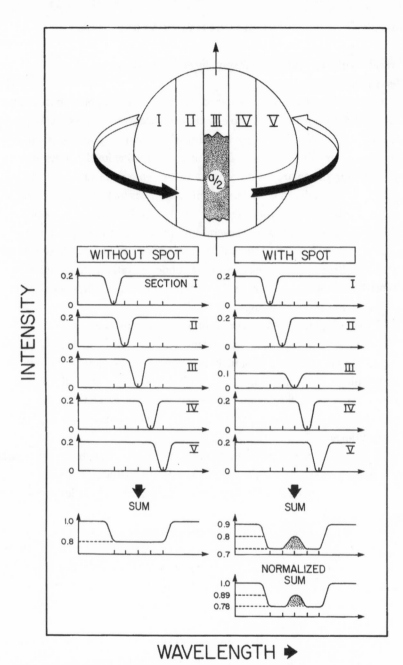

Fig. 9.1. The principle of Doppler imaging (see text). Each area on the rotating disk emits a narrow Doppler shifted line. The emissions combine to form the broadened line the observer sees. If a dark spot is present on the disk, a bump appears in this broad line and sweeps across the line profile as the spot rotates across the disk.

latitude. This star had no polar spot, however, unlike nearly all other RS Can Ven stars.

In 1997, Hatzes and Vogt analyzed eleven years of Doppler images of the star HR 1099, on which polar holes were first discovered. The same cool polar spot persisted for the full eleven years, while smaller spots came and went during the same time. Two long-lived spots emerged at low latitudes and migrated to join the polar spot, in contrast to the behavior of sunspots. The star rotated differentially in latitude, with high latitudes locked to the orbital period and lower latitudes rotating more slowly, also in contrast to the Sun.

In the years that followed, several groups applied the Doppler imaging technique to many spotted stars. Klaus Strassmeier and his colleagues at the Potsdam Observatory, as well as Swiss astronomer Svetlana Berdyugina, have devoted much of their careers to this work. In 2002, they independently summarized some of their findings.

Strassmeier noted that in a sample of sixty-five well-observed spotted stars, thirty-six have polar spots. (That certainly laid the controversy to rest.) Spots appear in stars of many types, including very young T Tauri stars; star systems similar to RS Can Ven and BY Draconis; solar-type stars; and red dwarfs. All are cool, with surface temperatures between 4,000 and 6,000 K. Many are giant or subgiant stars, meaning their masses and luminosities are larger than those of the Sun.

Star spots range in size up to 20 percent of the surface, but recent work has revealed clusters of smaller spots. A spot lives between a week and several years. (One persisted for sixteen years!) In general, smaller spots have shorter lives, as in the Sun.

The spots are found in roughly equal numbers at all latitudes on a star, including the rotational pole. In 1999 Steven Vogt, still active in the field, discovered another star in which the spots formed at low latitudes and migrated toward the poles, in contrast to the migration of sunspots toward the solar equator. The question arose as to whether spotted stars undergo a regular cycle similar to that of the Sun.

Recent research shows that the number of spots does vary cyclically on some stars, as revealed by changes in overall brightness. But unlike the Sun, the maximum number of starspots occurs when the star is at minimum brightness.

RS Can Ven is the perfect test case for a star with a cycle. Astronomers have been revisiting this bright binary frequently over the past 25 years, so that a nearly continuous record of its behavior now exists. In 2007 Svetlana Berdyugina analyzed the available data and extracted two spot cycles, of 16 years and 5.3 years, running in parallel.

The star's behavior is weird. Its spots flip between opposite hemispheres during the 5.3-year cycle. Moreover, the spots cluster at two opposite longitudes and migrate in latitude during the 16-year cycle. Spots at one active longitude migrate toward the pole, while those at the opposite longitude migrate toward the equator. If these regularities are confirmed, they will drive theorists mad!

Researchers naturally expected that these active stars, with their enormous starspots and violent flares, must possess magnetic fields and, presumably, active dynamos. But until the 1990s, they had no way to measure the fields; the critical magnetic signal (spectral line polarization) was completely lost in the much brighter unpolarized light of the star. These stars were not peculiar A stars, mind you, with kilogauss dipolar fields. Rather, they were cool giant G, K, and M stars, and although their spots were huge by solar standards, their magnetic signal was undetectable by the normal means. During the 1980s, an alternate method was devised that yielded some initial magnetic field strengths.

Richard Robinson, a solar scientist at the Sacramento Peak Observatory, revived a method previously used on Ap stars. It has the advantage of not requiring observations of polarized line profiles to detect the Zeeman effect. He suggested comparing the line profile of a magnetically sensitive line with the profile of a similar but magnetically insensitive line. The difference in their widths would yield the Zeeman splitting of the sensitive line and could provide estimates of both the field strength and the fraction of the surface covered by the field.

In 1984, Geoffrey Marcy at the Lick Observatory reported results with this method from twenty-nine G and K solar-type stars. He derived field strengths of 600 to 3,000 gauss, with filling factors as large as an extraordinary 89 percent. Steven Saar at Harvard University improved the method with calculations of spectral line formation and applied it to many types of stars. In the BY Draconis–type flare star EQ Virginis, for example, he derived a mean field strength of 2,500 gauss, covering a fractional area of 80 percent. These huge fractional areas raised a lot of eyebrows among researchers. After all, the method is subject to a variety of uncertainties. Astronomers hoped for something better. As we shall see, their hopes were soon realized.

BACK TO THE PECULIAR A STARS

Artie Hatzes got a bright idea. Why not apply the Doppler imaging technique to peculiar A stars? Instead of looking for starspots, he thought, look for the surface distribution of the elements that have anomalous abundances. So in 1989, Hatzes, Penrod, and Vogt collaborated in a study of the Ap star gamma2 Arietis, otherwise known as Mesarthim.

They obtained a series of spectra of an ionized silicon line at the Lick Observatory and looked for periodic variations in the strength of the line. They discovered two circular patches in which silicon is underabundant by a factor of 5. One spot coincided with the negative magnetic pole that Borra and Landstreet had identified previously, and was surrounded by a ring in which silicon is enhanced.

Here was a striking correlation. The authors suggested that depletion coincides with vertical magnetic field lines and enhancement with horizontal field lines. In accord with Michaud's theory, silicon ions might diffuse upward in the atmosphere of the star but might also accumulate where a horizontal field blocked their further rise. This idea was supported by current theoretical calculations. So it appeared that Michaud's theory of diffusion could be modified to take magnetic forces into account, an important advance.

This study opened the door to a new era of research on Ap stars. Many researchers began to investigate the spatial distribution of elements on their surfaces and how they relate to the shape and strength of magnetic fields. For instance, Artie Hatzes examined three Ap stars that were known to have anomalous chromium abundances. He discovered the same relationship between the distribution of chromium abundance and the orientation of magnetic field lines that he had seen for silicon in the star Mesarthim. Specifically, he saw patches of underabundance aligned with vertical fields and overabundance in horizontal fields.

Until the early 1990s, a researcher like Hatzes, who wanted to compare the surface distribution of an element with the magnetic field on the same star, had to rely on a previously derived oblique rotator model. These models assumed the field was dipolar, or at best, multipolar. They did not allow for any local variations of the field such as starspots.

This obstacle was removed in the early 1990s, when French astronomers Jean-François Donati and Meir Semel independently invented the technique of Zeeman-Doppler imaging. In essence the method consists of Doppler imaging in the circularly polarized light of one or more spectral lines. That differs from the simpler Doppler imaging, which was done only in the unpolarized light of a spectral line. The additional information coded in polarization allows an astronomer to construct coarse maps of the magnetic field, without any assumption that the field is smooth or dipolar.

The most important application of the method was to Ap stars. By choosing different spectral lines, it was now possible to map the distribution of different elements over a star's surface. Then one could compare the distribution of an element with the local strength and orientation of the magnetic field. Vast new opportunities were opened up with the advent of this clever technique.

Donati and Semel tried out their new method on the Ap star epsilon Ursae Majoris, known to the Arabs as Alioth. It is the fifth star in the tail of the Big Dipper. They were able to detect a relatively weak polar field of 186 gauss. But contrary to Hatzes's results for chromium, they found that iron ions concentrate at the magnetic poles. On another star, alpha2 Can Ven, iron seemed to collect preferentially at the magnetic equator. The picture was becoming rather confusing.

To explain these new developments, Georges Michaud collaborated in 1991 with Swiss astronomer Jean Babel in revising the old theory. They introduced ambipolar diffusion into the picture. In a vertical magnetic field, they showed, ambipolar diffusion would cause protons and electrons to drift upward and neutral hydrogen to drift downward. The protons would drag such ions as iron and chromium toward the poles. In a horizontal field, ambipolar diffusion would be suppressed and radiative diffusion would prevail. So the surface distribution of elements should depend on the orientation of the field. Voilá!

But as observations of the magnetic field and the surface distribution of abundances have steadily improved, the overall picture has become very complicated. In figure 9.2 we see corresponding maps of three elements and the magnetic field on the star alpha2 Can Ven. This famous Ap star is covered in patches that don't correspond very well with each other or to the local variations of the contorted magnetic field. Another star, HR 3831, shows patterns that are also poorly related to the field (fig. 9.3).

In a further improvement of the Zeeman-Doppler technique, independent observers like Oleg Kochukov at the Uppsala Astronomical Observatory in Sweden; Theresa Lueftinger at the Institute for Astronomy in Vienna; and George Wade at the University of Toronto were using all four polarization components of a spectral line to map surface distributions of elements and magnetic fields (note 9.3). With the higher resolution this method yields, stellar magnetic fields turn out to be much more complicated than first thought. And the correspondence of abundances and field geometry has become even weaker.

Obviously, some other factors help to determine the surface patterns. Two possibilities have arisen in the past five years. First, there is evidence that the chemical composition of an Ap star varies with depth in the atmosphere. Second, some Ap stars have been found to oscillate nonradially, like the Sun (note 9.4). Their surfaces are covered with patches that rise and fall periodically. Perhaps these motions help to generate the spotty surfaces of these stars. Time will tell. All we can say just now is that observations are outstripping theory by a far margin.

THE ORIGIN OF AP MAGNETIC FIELDS

Beginning with Babcock's discovery of the strong magnetic fields on Ap stars, theorists have debated about their origin. Their first thought was that some type of dynamo could be responsible, in analogy to the Sun. After all, Ap stars rotate differentially in latitude, and if some convection were present in the outer layers, it might be possible to construct a viable dynamo model.

The problem with that idea is that A stars are so hot that hydrogen is strongly ionized even near the surface. That means that in most of the volume of these stars, energy is transported to the surface by radiation, not by convection. The radiative zone is stable against the onset of convection. Without convection, the usual alpha process for transforming toroidal to poloidal fields would be absent and a solar type of dynamo wouldn't work. (See our discussion in chapter 3.)

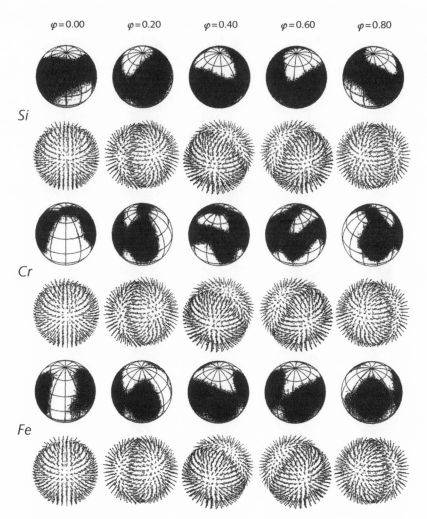

$\varphi = 0.00$ $\varphi = 0.20$ $\varphi = 0.40$ $\varphi = 0.60$ $\varphi = 0.80$

Si

Cr

Fe

Fig. 9.2. Vector magnetic fields and the distribution of silicon, chromium, and iron on the Ap star alpha2 Can Ven. As the star rotates (images from left to right), regions of enhanced abundance (white) and depleted abundance (dark) appear. There is little correspondence between elements or between an element's distribution and the orientation of the field. The letter φ signifies the fraction of a complete rotation of the star.

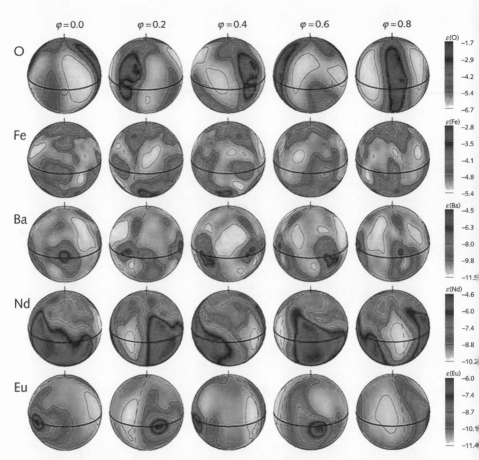

Fig. 9.3. More comparisons between the distributions of elements over the disk of HR 3831. In each row, you see the surface after the star has turned by 72 degrees. The letter φ signifies the fraction of a complete rotation of the star. The columns correspond to the different elements: oxygen; iron; barium; neodymium; europium.

Juri Toomre and his colleagues at the University of Colorado have recently reexamined the viability of a dynamo in an A star. They carried out three-dimensional simulations of convection at different rates of rotation and discovered that a seed field can be strongly amplified and reach several hundred gauss. Also, the magnetic patterns in the core are highly unsymmetrical and time-dependent. But whether these strong fields can migrate through the radiative zone to the surface is in question.

Twenty years ago Günther Rüdiger at the Astrophysical Institute in Potsdam proposed a type of dynamo suitable for a rapidly rotating A star. It turned out to have a predominantly quadrupolar field. As observations of a mainly dipolar field, slow rotation, and weak differential rotation of A stars accumulated, he realized that a new approach would be necessary.

Recently he has proposed a more complex form of dynamo. In his 3-D simulations, he prescribes a pattern of rigid rotation near the rotation axis and differential rotation farther from the axis. Such a pattern becomes unstable in the presence of a magnetic field and creates the helical turbulence that can supply the alpha effect. The question remains whether the model can be self-sustaining.

With such difficulties surrounding the dynamo, other theorists have considered an alternate explanation for the magnetism of A stars. Jon Braithwaite and Hendrik Spruit at the Max Planck Institute for Astrophysics, in Garsching, Germany, propose that the fields are fossils that were amplified as the star contracted out of the interstellar gas clouds.

Here the problem is whether the field would decay because of electric resistivity during the star's lifetime of a few billion years. The two men carried out numerical calculations of the evolution of a seed field to test the idea and learned that a predominantly toroidal field develops deep in the star, migrates to the surface, and transforms slowly and spontaneously into a poloidal field. Their ideas remain to be tested.

So, as in many situations in astrophysics, a phenomenon that seems simple at first turns out to be far more complicated on further examination. But we wouldn't want to solve all our puzzles too easily, would we?

COMPACT OBJECTS

On the night of January 31, 1862, Alvan Graham Clark was testing a newly finished telescope at the Dearborn Observatory in Evanston, Illinois. He and his father, Alvan Clark, were famous Boston opticians who had fabricated the lens of the telescope. To test the lens, they were looking at close double stars to see whether they could resolve them.

At one point in the evening, Alvan looked though the telescope at Sirius, the brightest star in the northern hemisphere. He was astonished to see a faint star some seven arcseconds from Sirius. No other telescope had been able to detect such a faint companion in the glare of such a bright star. He was delighted because it demonstrated the quality of his workmanship.

Eighteen years earlier, the German astronomer Friedrich Bessel had detected a puzzling wobble in the motion of Sirius. He predicted that the star possessed an unseen companion with a mass about equal to the Sun's. Clark had evidently found Bessel's predicted star. It was white in color and dim.

Then, in 1914, Walter Adams at the Mount Wilson Observatory determined that the companion (now called Sirius B) has a spectrum similar to a normal F star. Sir Arthur Eddington, the foremost astrophysicist of his time, assumed that the star's surface temperature is about 8,000 K and from its brightness and distance, calculated that the star has a diameter of only 37,000 km. It was smaller than Uranus! With such a small diameter, a solar-mass star would have a central density of 61,000 g/cm³, or 3,000 times the density of platinum (note 10.1), "which is absurd," he scoffed.

But when Adams confirmed that the spectrum of Sirius B (the first "white dwarf") is red-shifted by its enormous gravity, as Einstein's theory of relativity predicts, Eddington reluctantly admitted that the star was indeed as small and dense as the observations required. He still had reservations, however. "I do not see how a star which has once got into this compressed condition will ever get out of it," he wrote in his 1926 classic *The Internal Constitution of the Stars.*

Eddington recognized the basic problem with the star: its internal gas pressure could not balance its enormous gravity. How would the star avoid collapsing?

In 1926, Cambridge University physicist William Fowler supplied the answer. He applied the law of statistical mechanics recently announced by Enrico Fermi and Paul Dirac to the problem of white dwarfs. The law applies to particles like electrons and protons that are indistinguishable and that obey Pauli's exclusion principle (note 10.2). Fowler showed that electrons are so highly compressed in a white dwarf that they occupy all the available low-energy quantum states. Such a "degenerate" electron gas exerts a pressure with quite a different relation between temperature and density than ordinary gases. It is this pressure that prevents the dwarf from collapsing in its self-gravity.

A 20-year-old Indian physicist, named Subrahmanyan Chandrasekhar, added an amazing qualification to Fowler's result. In 1930 he was sailing to England to begin graduate study at Cambridge University. Chandra, as his friends called him, was a mathematical prodigy and a self-taught physicist. With plenty of spare time on the ship, he picked up the problem of the gravitational stability of a white dwarf. When he applied relativity to the degenerate state of electrons, he discovered a remarkable result: a degenerate star could not exceed a mass of 1.44 solar masses without collapsing. With any mass larger than this limit, the star's gravity would overcome the pressure of the degenerate electrons and catastrophe would follow.

Chandra later revised and extended his result. He was invited to present his work at a meeting of the Royal Astronomical Society in 1935. After he finished his talk, Eddington rose to criticize Chandra's application of relativity to quantum mechanics. He spoke with some authority, because he had once claimed to be the only one besides Einstein who actually understood relativity. And all through the 1930s, he severely criticized Chandra's mass limit.

Eddington was at the peak of his enormous reputation in these years, whereas Chandra was a modest newcomer 28 years younger. Their ongoing debate went beyond the usual bounds of scientific disagreement and became quite bitter. Paul Dirac and Niels Bohr, among others, came to Chandra's defense, but Eddington would not yield. Chandra had to suffer public ridicule from a senior figure but was finally proven correct. In 1983, Chandrasekhar and Fowler shared the Nobel Prize in Physics for their pioneering research. After Eddington died, Chandra wrote a gracious biography of the grand old man, touting his many contributions and avoiding any recrimination.

BIOGRAPHY OF A WHITE DWARF

More than 2,500 white dwarfs have been detected so far in the Milky Way. We now know that they represent the final stage in stellar evolution. A star of several solar masses generates its starlight by converting hydrogen to helium in its core. After several billion years, when the hydrogen is exhausted, the core of the star contracts and its atmosphere expands. The star becomes a red giant. Its central temperature rises, and it begins to convert helium into carbon and oxygen. After about a billion years, the helium fusion occurs only in a shell that surrounds an inert core of carbon and oxygen. The burning of helium is highly sensitive to fluctuations in temperature, and therefore the star may begin to pulsate in size and brightness: it may become a long-period variable star. With each pulsation, mass can be ejected from the low-gravity extended atmosphere. Alternatively, the star may eject mass continuously in a stellar wind that is driven in part by the ultraviolet light from the radiating surface. After the helium is exhausted, the star must contract further to raise its central temperature in order to convert carbon to neon. It can only do this if it has sufficient mass. If it lacks the minimum mass, it sheds its outer layers in a powerful wind to form a shell-like planetary nebula. The remaining core, composed of carbon and oxygen, is a white dwarf.

White dwarfs are born hot but have no internal source of energy. As they radiate their heat, they cool slowly, over billions of years. Gradually they fade out and change in color from white to orange to red. In principle they could wink out entirely and become invisible. (Some theorists think that these black dwarfs could be part of the dark matter of the galaxy.)

White dwarfs have hot atmospheres, a few tens of meters thick, which are composed largely of hydrogen and helium. In the powerful gravity of the star, the two elements are separated, with helium pulled down to the bottom of the atmosphere and hydrogen floating on the top. The atmospheres range in temperatures from as low as 5,500 K to at least 100,000 K, and their spectra vary correspondingly. Thus, some show strong spectral lines of hydrogen; others have lines of neutral or ionized helium. Although the atmospheres are thin, their opacity determines how slowly the dwarf cools.

Among all the exotic properties of a white dwarf, none is more impressive than its magnetic field. We turn now to this topic.

SMALL BUT MIGHTY

In 1970, James Kemp made a remarkable prediction that turned out to have wide applications in astrophysics. He was a professor of experimental solid-state physics at the University of Oregon. When I met him in 1973 he was a tall, shy man who seemed always to be on the point of leaving the room on an important errand, like the White Rabbit in *Alice in Wonderland*.

Kemp had been studying the optical properties of solids, like magnesium oxide, in magnetic fields as strong as 24 kilogauss. Magnesium oxide is a "birefringent" material; it transmits different polarized states of light at different speeds. In the course of his laboratory work, Kemp discovered an unusual property of hot magnetized solids: their white light could be partially polarized. That led him to predict that in a very strong magnetic field, even a hot gas would emit a small fraction of circularly polarized white light.

Kemp visited my colleagues and me at the University of Hawaii in 1973 and tried to convince us that his physics was sound. We had learned in standard optics courses that the smooth continuous spectrum of a thermal source like a hot gas is unpolarized. Only spectral lines could be polarized in a magnetic field. Kemp explained that in a very strong field, free electrons emit left-circular and right-circular light in slightly unequal amounts. A net polarization would remain. He suggested that a white dwarf might exhibit the effect and was eager to confirm it.

Roger Angel at the University of Texas and John Landstreet at the University of Western Ontario jumped at this idea. They immediately searched for polarized emission in several white dwarfs and reported unhappily in 1970 that they could find none. At the same time, Kemp sprinted to the Pine Mountain Observatory of the University of Oregon, and using its 24-inch telescope with a clever polarization detector he had invented, discovered the first white dwarf to show the effect. It was a dim little fellow, barely visible in the telescope.

Then Kemp, his laboratory colleague John Swedlund, Angel, and Landstreet went on to the Kitt Peak National Observatory and confirmed the discovery with a 36-inch telescope. When they applied Kemp's theory to the data, they could hardly believe their eyes. The dwarf had a field of 12 megagauss!

In the 1970s, Roger Angel and his colleagues embarked on an intensive campaign to measure the magnetic fields of isolated white dwarfs. They achieved a breakthrough at the 200-inch telescope in 1974, when they measured the shape and circular polarization of hydrogen spectral lines in the dwarf G90. By matching the observations with synthetic spectra, calculated for different field strengths, they determined its field strength as approximately 5 megagauss. They estimated that 20 percent of the dwarf's surface was covered with this field.

Many other observers applied this method in the 1980s and 1990s. By 2000, sixty-five isolated magnetic dwarfs were known, with fields ranging from less than 100,000 gauss to 500 megagauss. The most common field strength is about 16 megagauss. Magnetic white dwarfs are relatively rare, however. Only about 5 to 10 percent of isolated dwarfs have detectable fields. Among cataclysmic variables, they are more common, at about 25 percent.

A white dwarf with the unmemorable name of PG 1031 + 234 holds the record for the strongest magnetic fields, a whopping 500-megagauss polar field and a 1,000-megagauss starspot. Compare that to the maximum magnetic fields available in the laboratory (about 450,000 gauss, or double that for short periods) and you will understand why astrophysicists are eager to explore the behavior of matter in white dwarfs.

All the early determinations of field strength were based on the assumption that the shape of the field is dipolar or perhaps a decentered dipole. That was the simplest assumption and the only one that the precision of the observations justified. But as the quality of spectroscopic and polarimetric observations improved, more complicated shapes were detectable.

A group of German scientists based at the universities of Göttingen, Heidelberg, Kiel, and Tübingen has pushed the method to the limit. In 2002 they announced that they had computed over 20,000 synthetic polarized spectra, for field strengths between 10 and 200 megagauss. Then they began to observe white dwarfs with the 8-meter telescopes at the European Southern Observatory in Chile. They obtained spectra during the rotation of a dwarf and used a clever optimization technique to fit the data to a variety of magnetic models. They call their method Zeeman tomography, in analogy to computed axial tomography in medical diagnosis.

In 2005 they reported their results for a magnetic dwarf that rotates in 2.7 hours. The field turns out to be primarily quadrupolar, with additional dipolar and octupolar components. The predominant field strength is 16 megagauss, but patches of 10 and also 75 megagauss are present. These researchers have high hopes of being able to map the surfaces of magnetic white dwarfs in the same way that Ap stars have been charted.

How are these monster magnetic fields generated? Several ideas have been proposed. The favorite seems to be that the fields are fossils, left over from the formation of the white dwarf. If a star could retain all its magnetic flux through this violent event, the field strength would increase in proportion to the square of the radius. For example, an Ap star with a surface field of, say, 1,000 gauss could end as a white dwarf with a field of 10 megagauss. But could the star actually retain its magnetic flux during the violent birth of a white dwarf?

Another alternative is that the original hot star had a convection zone and a working dynamo before it erupted. And yet another idea is that a dynamo of some sort might be working at the present time in a white dwarf, either in the liquid core or just under the surface. But theorists are far from understanding the details of such a scenario. We must wait for some clarity.

AN UNEASY PARTNERSHIP

White dwarfs can cause spectacular outbursts of visible light and x-rays when they are paired with an ordinary star in a binary. Such a combination is called a cataclysmic variable (CV) and the name is really appropriate. A

CV can suddenly brighten by a factor of a million, in a so-called classical nova. These events are not supernovas, mind you; they release a million times less energy and have an entirely different origin.

In a classical nova, the dwarf's gravity tears a chunk of hydrogen gas off its companion and the hydrogen crashes onto the dwarf's surface. Temperatures of more than 100 million kelvin are produced, sufficient to ignite a thermonuclear reaction. The resulting explosion can release 10^{31} kilowatt-hours and a powerful pulse of x-rays. Although another nova may not occur in the same CV for another 10,000 years, an average of thirty novas are estimated to occur per year in our galaxy.

CVs can also produce a different kind of eruption, which occurs much more frequently. It's about a hundred times weaker than a classical nova, may last for one or two weeks, and may recur every hundred days or so. These events arise in an accretion disk that revolves around the equator of the white dwarf. The disk is fed by material that continuously spirals in from the dwarf's companion. Occasionally the disk overfills and dumps hydrogen onto the equator of the white dwarf. Then the gravitational energy that the gas gains in its free fall supplies the energy for the resulting nova (note 10.3).

If the white dwarf possesses a magnetic field of several tens of megagauss, the mechanics of the nova are quite different. In so-called polar CVs, such as the binary AM Hercules, the dwarf's strong field prevents the formation of an accretion disk. The field also locks the dwarf's rotation period to the orbital period.

The dwarf pulls plasma off its companion star, as in nonmagnetic binaries. But now the dwarf's magnetic field deflects the plasma and channels it to fall on the magnetic poles. The plasma descends in vertical columns or curtains and creates a strong shock above the dwarf's surface. The shock heats the plasma to hundreds of millions of kelvin, and radiates hard x-rays.

An "intermediate polar" CV, like DQ Hercules, also contains a magnetic white dwarf, but its field strength is more moderate, perhaps only a few megagauss. These CVs do possess an accretion disk, which overflows occasionally. But as in the polar CVs, the infalling gas is guided by the dwarf's magnetic field to land at the magnetic poles.

NEUTRONS, CHEEK TO CHEEK

White dwarfs are strange, but neutron stars are almost beyond belief. Imagine squeezing a Sun-like star until all the nuclei of its atoms are touching, with no spaces in between. Simple arithmetic shows that the object would have a diameter of about 10 or 20 km and each cubic centimeter would weigh a billion tons here on Earth (note 10.4). The object's self-gravity would be sufficient to hold itself together without our aid. We would have created a neutron star.

Neutron stars were predicted to exist before they were actually discovered. In 1934 Walter Baade and Fritz Zwicky, astronomers at Mount Wilson Observatory, announced their discovery of a new type of nova, which they named "supernova." Baade was the courteous, meticulous astronomer of this famous pair. He is perhaps best known for discovering that the light of the Crab Nebula is completely polarized. Zwicky was the gruff, egocentric genius, who had little time for the small minds at Caltech. Long before anyone else, he discovered dark matter in clusters of galaxies and proposed the possibility of gravitational lensing. Not only a fount of innovative ideas, he was also a hard-working observer, who discovered more than a hundred supernovas.

This odd couple calculated that supernovas release a thousand times the energy of the more common novas and that the total energy loss is equivalent to the annihilation of a substantial fraction of a normal star's mass. For the source of the supernova's energy, they made a radical suggestion:

> With all reserve we advance the view that a supernova represents the transition of an ordinary star into a neutron star, consisting mainly of neutrons. Such a star may possess a very small radius and an extremely high density. As neutrons can be packed much more closely than ordinary nuclei and electrons, the "gravitational packing energy" in a cold neutron star may become very large, and, under certain circumstances, may far exceed the ordinary nuclear packing fractions. A neutron star would therefore represent the most stable configuration of matter as such. (*Publications of the National Academy of Science* 20, no. 5, 259, 1934)

In essence, Baade and Zwicky proposed that a supernova derives its energy from the gravitational energy that is released when the core of a massive star can no longer support itself by its thermal pressure and collapses. In a burst of inspiration, they also suggested that supernovas are the sources of the high-energy cosmic rays that we observe in the galaxy. Their assertions were based on fragmentary observations and rough calculations, but they were later proven correct.

Baade and Zwicky's concept of a neutron star lay dormant for thirty years. Then in 1967, a young student practically fell over one.

LITTLE GREEN MEN

Jocelyn Bell was one of Antony Hewish's graduate students at Cambridge University. Hewish and his students had built a radio telescope to study quasars. While analyzing the data, Bell uncovered a strange variable source that emitted a sharp pulse every 1.33 seconds. At first Hewish and his team speculated that they had detected an extraterrestrial civilization (Little Green Men) but after finding three more such sources, they decided they were natural. It was one of the great discoveries of the twentieth century (note 10.5).

What could they be? How could one explain the intense emission and the absolutely steady pulse rate? Some researchers thought the objects might be spinning white dwarfs, but that idea died when they realized that centrifugal forces would tear apart a dwarf.

In 1968, Thomas Gold at Cornell University proposed a solution that has proved to be correct. Gold was a colleague of Fred Hoyle and Hermann Bondi at Cambridge University in the 1940s. These three brilliant physicists proposed the steady-state theory of the expanding universe. It had no beginning and would have no end. It would always look the same, despite its expansion. To fill the gaps that opened between the galaxies, the trio postulated the spontaneous creation of new matter. Their theory was disproved and George Gamow's Big Bang was confirmed when the cosmic microwave background was discovered in 1965.

Gold enjoyed provoking scientific controversies and was fearless in proposing seemingly outrageous explanations for new discoveries. Unavoidably, he was the author of several bold predictions that proved incorrect. He predicted, for example, that the first Lunar Lander would sink into meters of black dust and that the interior of the Earth contains an inexhaustible supply of methane fuel. He will always be remembered, though, for his model for Hewish's pulsating radio sources.

In 1968, Gold proposed that the sources were rapidly rotating neutron stars that possessed extremely powerful magnetic fields. The radio radiation, he asserted, is emitted in a narrow beam of high intensity from each magnetic pole. In general the magnetic axis is tipped with respect to the rotation axis. Consequently, as the star rotates, each beam sweeps by the Earth like the beam from a lighthouse, and we detect sharp pulses. The spinning of the star also accounts for the steady pulse rate. But in the long run, we should expect the pulse rate to decrease, because the star's rotational energy is the source of the energy it loses as radiation.

Gold's model was based on estimates of the diameter and magnetic field of a neutron star. He could easily estimate the diameter (note 10.4), which turned out to be an astounding 10 or 20 km. Then, assuming all the magnetic flux of the original star was retained during the collapse to the neutron star, he could estimate the final field strength. It would increase in proportion to the ratio of the squares of initial and final diameters, a factor of 10^{10}. So if the original field were 1 gauss, like the Sun, the final field strength would be a staggering 10,000 megagauss.

But this estimate of the field strength was only a lower limit. Gold proposed a simple model to explain the observed intensity of the radio radiation. He reasoned that a rapidly rotating magnetic field would induce electrical currents that would flow along the field lines emerging from the magnetic poles. The currents would generate radio waves that would propagate outward along the polar field lines. But to account for the intensity of the radiation, he required a magnetic field strength of 10^{12} gauss, or a million megagauss. Such fields were orders of magnitude larger than any detected in white dwarfs.

Gold asked to announce his model at a meeting of the American Physical Society in May 1968 but was refused time. He quoted the organizers saying, "The suggestion was so outlandish that if this were admitted, there would be no end to the number of suggestions that equally have to be allowed" (*Nature* 218, 731–32, 1968). Gold was able to publish his paper in the journal *Nature*, however, and the rest is history. A journalist contracted the words *pulsating* and *star* and coined the name "pulsar."

Later that year, two radio astronomers at the National Radio Astronomy Observatory detected two pulsating radio sources in the vicinity of the Crab

Nebula. The pulses were sporadic and showed no signs of periodicity. But because the nebula is a well-known remnant of a supernova that Chinese astronomers saw in the year 1054, there was a possibility that at least one of these sources could be similar to Hewish's pulsar.

John Comella and his colleagues at the Arecibo Radio Observatory in Puerto Rico decided to take a look. The observatory's 1,000-foot telescope allowed them to focus tightly on the sources. They determined that one of them was indeed a pulsar and that it is spinning at an incredible thirty times a second.

A year later Comella and D. W. Richards were able to report that the Crab pulsar is slowing down. Its period of rotation is increasing at about 1 part in 2,400 per year. This critical observation confirmed Gold's prediction that a pulsar's rotational energy slowly leaks away as radiation, and it went a long way toward confirming his basic model. Moreover, the rate at which the pulsar is slowing was consistent with a magnetic field of 10^{12} gauss.

In 1968, the Crab pulsar was known to pulsate only at radio wavelengths of a few meters. Then, within one year, astronomers discovered that this amazing object also pulses in hard and soft x-rays, visible light, and infrared light. In 1972, pulses of gamma rays were discovered, with photon energies up to 10^{12} electron volts.

When the total power radiated by the pulsar was added up, a serious problem emerged. The total amount of energy the pulsar loses per second could be determined from the rate at which it slows down. When the measurements of the radiation losses were compared with this total power loss, a big shortfall was found. The pulsar's radio radiation accounted for only a millionth of the total power lost, and the emission in x-rays and gamma rays accounted for at most a tenth of the power. So the pulsar was losing energy in an invisible form. What could it be?

The answer emerged when the Crab Nebula's radiation losses were totaled. In 1968, Robert Haymes and his colleagues at Rice University measured the gamma ray spectrum of the nebula from a high-altitude balloon. They detected gamma rays with photon energy up to $500\,\text{keV}$. But more importantly they demonstrated that the gamma ray spectrum joins smoothly to the known spectra of radio waves and visible light. They concluded all the light from the nebula is synchrotron radiation, produced by energetic electrons that spiral around its magnetic lines of force.

By adding up all the radiative losses of the nebula, Haymes and his team estimated the total equals 20,000 times the power output of the Sun. Moreover, these losses are approximately equal to the loss of rotational energy by the Crab pulsar. The unavoidable conclusion was that the Crab pulsar powers the entire nebula by continuously injecting electrons with energy up to at least as high as 10^8 MeV. The question then is "How are these electrons accelerated?"

Peter Goldreich and William Julian at Caltech offered an explanation in 1969. They constructed a model for a pulsar's field and its interaction with its surroundings. First, they showed that a pulsar must have a magnetosphere filled with dense plasma. Their reason was that rotation of the field induces an electric field at the stellar surface, which pulls charged particles out of the surface despite the huge gravitational force.

In their model (fig. 10.1) the field has three divisions. Near the star, the field is a dipolar magnetosphere and co-rotates rigidly with the star, out to a radius where the speed of rotation approaches the speed of light. Beyond that radius, the field lines at the poles open out into a wind zone that extends to a boundary zone inside the supernova shell. The "open" lines of force from the poles extend through the wind zone and close into a toroidal field near the supernova shell. As the field winds up, it exerts a braking torque on the star, causing it to spin down.

Goldreich and Julian suggested that electrons and protons leave the magnetosphere separately, on different field lines. The rotation of the field causes the charges to whip out along the field lines like beads on a wire. The particles are accelerated to relativistic energy and carry off most of the energy the star loses by spinning down. They fill the nebula behind the supernova shell and release their energy in synchrotron radiation.

Here, then, was a first cut at a model to explain the source of the synchrotron radiation of the Crab Nebula. Most of the energy lost by the spinning neutron stars escapes as fast charged particles not radiation, just as the observations require. Goldreich and Julian predicted the rate of the pulsar spindown; the maximum energy of the particles; and the strength of the magnetic field in the nebula. But their prediction of the rate at which electrons are injected into the nebula fell short by a factor of 1,000.

F. Curtis Michel at Rice University introduced an alternative model in which the density of charged particles in the pulsar magnetosphere is so

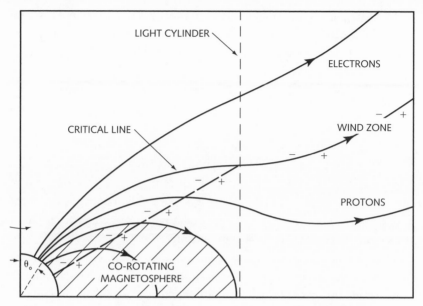

Fig. 10.1. An early model of a pulsar. Electrons and protons reach the speed of light along separate lines of force that emerge from the poles of the neutron star. The charged particles carry off most of the energy that the star loses by spinning down.

large that they can be treated as plasma. Then he could use magnetohydrodynamics theory to model the pulsar field as a force-free field. He formulated the problem of the spin-down and energy flow in terms of a nasty equation. For thirty years, neither he nor anyone else could find a physically acceptable solution to this equation. Finally in 1999, the nut was cracked. Since then, many researchers have produced detailed numerical examples of the solution. Figure 10.2 shows a pulsar field that Anatoly Spitkovsky at Stanford University calculated in 2007.

Notice that within a few radii of the star, the field is not sharply focused near the magnetic poles, as one might expect from the lighthouse analogy. And yet pulsar radio pulses are quite brief: they last for only a small percentage of the star's rotation period. So the narrow beam must develop in the field at greater distances, where the toroidal component begins to predominate.

Even with a realistic magnetic field model in hand, researchers have many questions regarding a pulsar's intrinsic x-ray and gamma ray emissions. Pulsars that rotate in tens of milliseconds emit only pulsed x-rays, with a non-

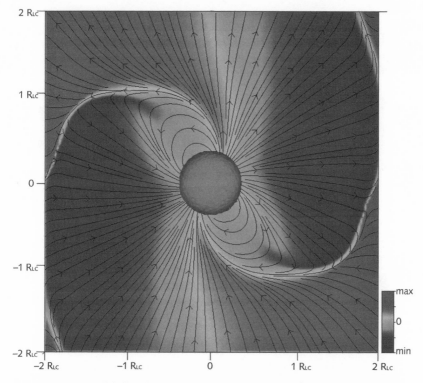

Fig. 10.2. A recent model of a pulsar magnetic field. We are looking down on the unique plane that contains both the rotation and the magnetic axes, which are separated by 60 degrees. The lines of force in the light gray area tilt out of the plane toward us, while those in the dark gray tilt away from us. The overall pattern is an Archimedes spiral.

thermal (synchrotron) spectrum. Pulsars that rotate in only a few milliseconds emit thermal x-rays continuously, as well as pulsed nonthermal x-rays. Hot spots on the stellar surface can account for the steady thermal emission, but the pulses are still puzzling.

Part of the problem is that a pulsar's pulses are not all the same. Their profiles (of intensity vs. time) vary from one pulse to the next. One interpretation is that the currents that generate both the radio and x-ray pulses are themselves variable and shift around the pulsar magnetosphere randomly. In addition some pulsars, like the Crab, occasionally emit "giant" radio pulses. In 2003, observers detected isolated, highly polarized radio bursts only 2 nanoseconds long. These variations will keep theorists busy.

The physics of extremely strong magnetic fields is another frontier. The early pulsar models were consistent with field strengths of about 10^{12} gauss. But in 2003, observers at the Parkes Radio Telescope detected several pulsars with fields of 10^{14} gauss, as deduced from their spin-down rates. In such strong fields, quantum mechanical effects become important. Pairs of electrons and positrons can be created spontaneously, for instance, and could supply most of the charged particles. Quantum effects can also affect the production of synchrotron radiation and the Compton scattering of gamma rays, in which a photon collides with a particle and transfers energy to it. So theorists are busy crafting models that could be tested against new observations.

Pulsars grow old and die. We know now that only a few young pulsars, like the Crab, that rotate in much less than a second, produce gamma rays. As a pulsar ages and its spin rate declines, first its gamma rays fade and then its x-ray emission. At an age of about 100 million years, even its radio pulses are predicted to disappear. Just how this aging process occurs will challenge the experts for some time.

A DIFFERENT KIND OF BEAST

On March 5, 1979, nine spacecraft spaced around the solar system detected a gamma ray burst one hundred times more intense than any recorded before. The burst lasted a mere 0.2 seconds and was followed by a train of hard x-ray pulses, 8 seconds apart. The pulses slowly died out after about 200 seconds. Then the source turned off. No radio pulses, no x-rays. Nothing. Seasoned observers like Thomas Cline at the NASA Goddard Space Flight Center had never witnessed anything like it.

But the best was yet to come. Cline and colleagues pinned down the location of the source to a supernova shell in the Large Magellanic Cloud, a satellite galaxy of our Milky Way that lies at a distance of 168,000 light-years. That meant that the huge burst Cline and company had detected was a mere whisper of a stupendous event that occurred far away. A quick calculation showed that in 0.2 seconds the source had released more energy than the Sun radiates in a thousand years.

Astronomers were mystified. Because the source lay at the edge of a supernova remnant, they speculated that it could be a rotating neutron star.

But if so, its 8-second rotation period would mark it as much older than the supernova in which it was embedded. Moreover, its gamma ray photons were much "softer" (less energetic) than, say, the Crab pulsar's.

The gamma ray pulsations were also unusual. Cline was well acquainted with "gamma ray bursters," which explode once and never again from the same point in the sky (note 10.6). He also was familiar with x-ray binaries that pulsated continuously. But this transient pulsating source of 1979 resembled neither of these objects.

Many explanations were proposed, some quite speculative. Perhaps a small asteroid had crashed into a neutron star or the interior of a neutron star had suffered a catastrophic upheaval. However, none of these ideas could explain the enormous burst of energy, and none gained any support. The theorists were stymied.

Then in 1986, astronomers at a conference in France compared observations and discovered that two similar sources had erupted on January 7 and March 14, 1979. Both were located in the Milky Way, both were associated with supernova remnants. Like the March 5 source, they emitted a short, intense burst followed by periodic pulsations. These three sources were given a name, "soft gamma repeaters" (SGR), but a convincing explanation for them still seemed far off.

During the 1980s, the January 7 source (labeled SGR 1806-20) flared up repeatedly, each time with its characteristic initial burst and periodic pulsations. In contrast the March 14 source (SGR 1900 + 14) remained dormant for two decades. There was no explanation for their eccentric behavior.

Then a wild idea appeared in 1992. Robert Duncan, a postdoc at Princeton University, and Christopher Thompson, a Princeton student, had wondered just how strong a magnetic field a neutron star might be born with. The conventional wisdom decreed that, as a neutron star collapsed, it simply compressed the magnetic flux of its massive parent star. The result, typically 10^{12} gauss, was consistent with the decay rates of pulsar spins. But suppose a neutron star could generate new magnetic flux during its birth. Suppose, in fact, that a dynamo was possible. What then?

Duncan and Thompson based their theory on published simulations of the formation of a neutron star. This was highly speculative stuff, but the calculations showed that the interior of the star was churning violently in the first few seconds after gravitational collapse. The star was born hot and was cooling rapidly as convection carried heat to its surface.

The two young scientists argued that the combination of a strong initial field and vigorous convection could constitute an alpha-omega dynamo. They predicted that if a neutron star was born spinning faster than 1,000 rotations per second, then its field could build up to as much as 10^{17} gauss. They coined the name "magnetar" for such a highly magnetized star.

After a few seconds of cooling, convection would turn off and the dynamo would also stop. The star would retain its strong field, but magnetic torques would continue to spin down the star, releasing its rotational energy as radiation and a magnetized wind. They estimated that the rotation period could increase to as much as 10 seconds in as little as 10,000 years.

Here was a scenario that could explain the presence of an apparently old (i.e., slowly rotating) neutron star at the center of a relatively young supernova shell. Duncan and Thompson suggested that the SGR observed on March 5, 1979, could be a magnetar. When they applied their theory to this event, they derived a dipolar field strength of about 6×10^{14} gauss, at least one hundred times larger than "normal" pulsar fields.

Other researchers were skeptical, to put it mildly. This theory seemed even more speculative than those that preceded it. But Thompson and Duncan were not discouraged. In a series of papers between 1992 and 1996, they continued to extend and quantify their basic theory. In the process they added even more radical ideas to the grand structure they were building. They would have a wild ride in the coming decade, as new observations appeared.

To make their magnetar proposal plausible, they proposed a mechanism to account for the dramatic gamma ray spike of March 5, 1979. They hypothesized that twisted magnetic flux in the interior of a magnetar would diffuse outward into the magnetosphere. As the magnetic field emerged, it would crack and scatter pieces of the solid crust of the star. The sudden rupture would cause a transient in the field and produce a soft gamma burst.

The twisted field would move out into the magnetosphere, where its energy would be stored. The twist would also increase the magnetic braking torque, so that the star's rotation would slow more rapidly. At some point, the twist would become unstable and would release its stored energy. Some of the energy would be detectable as a giant gamma ray burst; the rest would appear in two parts: an expanding fireball and a trapped cloud of plasma.

To account for the soft gamma pulsations that follow the initial burst, Duncan and Thompson claimed that the trapped plasma heats spots on the stellar surface. As the star rotates, the spots swing into view and produce the pulsations. After all the fireworks die out, the magnetar continues to glow in soft gamma rays, but it rotates too slowly to generate radio waves, despite its powerful field.

Their main point in all this is that soon after its birth, the SGR powers all its emissions and magnetized winds with the decay of its magnetic field, not by the loss of rotational energy.

We fast-forward to November 5, 1996. The source SGR 1806-20, which first flared in January 1979, was active again. Gamma ray detectors on the Compton Gamma Ray Observatory (CGRO) and on the Rossi X-ray Timing Explorer (RXTE) were nearly blown off-scale. Chryssa Kouveliotou, a researcher at NASA's Marshall Space Flight Center, had been assigned time on the RXTE for a different program. When news of the flare arrived, she decided to seize this opportunity. The RXTE was ideal for her purpose. It had been launched in 1995 to search the sky for variable x-ray sources. Its instruments could measure x-rays as hard as 250 keV, with variations as short as microseconds and as long as months.

Kouveliotou and her international team were lucky. From November 5 to 18, they recorded several hours of sharp x-ray pulses that were spaced 7.5 seconds apart but were bunched into discrete packets. If this SGR was a neutron star, it was rotating very slowly indeed. The Japanese satellite ASCA (Advanced Satellite for Cosmology and Astrophysics) had observed the same source during its quiescent phase in 1993 and had also recorded pulses at 7.5-second intervals. So at first it appeared that the team had merely confirmed a previous result. But they had to be sure.

Kouveliotou's team spent the next eighteen months reanalyzing both their data and those of the Japanese. Finally, in May 1998 they concluded that the two pulse periods differed by 0.008 seconds. If the source was a neutron star, its rotation period had decreased by this amount in only four years. That would imply an enormous loss of energy, despite a slow rotation. That, in turn, would imply the presence of an extraordinary magnetic field strength. When the team applied the standard theory of pulsar spin-down, they derived a dipole field of 8×10^{14} gauss. This was precisely in the range that Duncan and Thompson had predicted. The conclusion? This SGR could be a magnetar.

Other magnetar candidates appeared soon afterward. On August 27, 1998, the source SGR 1900 + 14, which had been dormant since March 1979, burst out with a gamma ray flare that broke all records for maximum energy. A train of 5.2-second oscillations followed. Kouveliotou, collaborating with a stellar cast of scientists, determined that this source was also slowing down rapidly. From the period and the rate of its decline, they estimated a magnetic field of 5×10^{14} gauss.

Up to this point, support for ultrastrong magnetic fields was based primarily on the spin-down of the repeaters. Alaa Ibrahim and his colleagues at the NASA Goddard Space Flight Center felt that independent confirmation was needed. So they sifted through the RXTE records for the November 1996 period of activity of SGR 1806-20 and found a jewel. A narrow absorption line appeared in the x-ray spectrum at 5 keV.

The data were noisy, but the line did appear in many small bursts. Ibrahim and friends at the center combined the data to obtain a precise line profile. They then identified this line with the absorption of cyclotron radiation of a proton in a strong magnetic field. Such absorption features had been seen previously in the spectra of gamma ray bursters. When they applied the standard theory of cyclotron absorption, they derived a magnetic field of about 10^{15} gauss. The magnetar model had another leg to stand on.

Since 1998, the study of SGRs has grown into a minor industry among astrophysicists, although only four are positively known. Several international teams are collaborating in an Interplanetary Network of five x-ray observatories (note 10.7), and theorists are also very active. The evidence that SGRs are magnetars has grown ever stronger.

Even the more controversial aspects of the Thompson-Duncan scenario have become plausible. One of the team's predictions concerns the cracking of the neutron star crust and its influence on the gamma ray bursts. They speculated that the emergence of twisted magnetic flux from the interior would fracture the crust. The release of energy from the stressed crust would power many small bursts and also contribute to the rare giant bursts.

In 1999, a team of scientists at the NASA Goddard Space Flight Center found evidence that was consistent with this picture. They analyzed the intensity of random bursts that SGR 1900 + 14 emitted during its 1998–1999 year of activity. They determined that the number of large and small bursts has the same distribution as earthquakes. The implication was that the crust powers the bursts, as in the magnetar scenario. The group concluded that the source of the energy was very likely the interior of the star and was "plausibly magnetic."

This same research group used data from the RXTE to determine the spin-down rate of SGR 1900 + 14. The star was slowing down at a constant rate before June 1998. Between June and the August 27, 1998, giant flare, the star's spin-down rate doubled. That meant the magnetic torque on the star had increased. And to account for the size of the change, Thompson and Duncan calculated that the star's magnetic field had to exceed 10^{14} gauss. Here was another point in favor of their model.

The observational picture that has emerged is as follows. An SGR can remain dormant for years, then wake up suddenly to emit hard x-ray bursts for weeks or months. A typical burst lasts about a tenth of a second and contains as much energy as the Sun emits in a year. Perhaps once in fifty years, the source may emit a giant flare, a thousand times more energetic and lasting about five minutes. Then the source may relapse into a quiescent state, continuously emitting as much energy in x-rays as the Sun emits in visible light.

On December 27, 2004, SGR 1806-20 produced the brightest extrasolar gamma ray burst ever detected, surpassing the brightness of the full Moon for 0.2 seconds. The total energy it released in five minutes equaled the total output of the Sun in 500,000 years. This single event will provide enough data to keep researchers busy for years, testing the magnetar model.

As an example, Peter Woods and his colleagues at the National Space Science and Technology Center, in Huntsville, Alabama, obtained observations of the event from the Chandra X-Ray Observatory and the RXTE. They were also able to compile a history of the source before, during, and after the dramatic burst. Their most interesting results concern the spin-down of the neutron star.

The star's rotation period was slowing at a constant rate of 2.6 ms per year between 1993 and 2000. Then in 2000, the spin-down rate suddenly jumped up by a factor of 6, although no increase in burst activity, persistent emission, or pulse profile was observed. This large spin-down rate persisted until July 2004, when it increased slightly. Therefore, the torque on the star reached a maximum five months before the giant flare in December. After the flare the spin-down rate (i.e., the torque) returned rapidly to the low level of 1993–2000.

So far Thompson and Duncan haven't attempted to compare these results with their model, but I expect they will.

Other groups are searching through their data for 1994 to find new ways to test the magnetar model. For instance, a large team from the Imperial College of London was observing the Earth's magnetosphere with satellites Cluster and Double Star TC-2 (see chapter 6). They were lucky to observe the first few milliseconds of the giant burst. The intensity of the burst rose in steps, at three different rates of increase. The group concluded that the rates of rise are consistent with the Thompson-Duncan idea of a neutron star's crust cracking.

Another large team, observing with the Ramaty High Energy Solar Spectroscopic Imager (RHESSI), obtained evidence in support of an unwinding twist in the magnetosphere during the giant burst. And yet another group, led by Joseph Gelfand, recorded the fading of an afterglow at radio wavelengths. Analysis of this glow has yielded a detailed model of the outflow, the formation of shocks, and much more.

This field of research is moving almost too fast to follow. The old data are still being mined fruitfully, and no doubt new observations will soon provide some surprises. So far the magnetar model seems to be holding up very well. Christopher Thompson and Robert Duncan have also prospered. Thompson is now a professor at the Canadian Institute for Theoretical Physics in Toronto and has received several prestigious awards for his creativity. Duncan has been appointed senior research associate at the University of Texas at Austin.

BLACK MAGIC

Everyone has heard or read about black holes. According to popular accounts, they are almost mythical objects that suck up everything around them: stars, gas, or dust. Their gravity is so powerful that even light rays cannot escape them. They are, if you can believe it, almost the size of a mathematical point but may contain the mass of many Suns.

So naturally you might expect that, in analogy to neutron stars, black holes would have magnetic fields of absolutely mind-boggling strength. Not so, theorists tell us. Black holes have only three physical properties: mass, angular momentum, and electric charge. All else is conjecture. But powerful magnetic fields do exist just outside a hole, and as we shall see, they help to prepare the meals the hole eats.

Karl Schwartzschild was the first to point to the possible existence of a black hole. In 1916 this German astrophysicist obtained a solution to Einstein's equations of general relativity, which describe the gravitational field in the vicinity of a mass. He determined that, at a critical radius from a massive object (the event horizon), a body would need to move at the speed of light to escape the gravity of the object. Because special relativity limits the speed of any mass to the speed of light, nothing, not even light, could escape from radii smaller than the event horizon. Anything within the event horizon falls into a point-like "singularity" of infinite density.

In 1939, J. Robert Oppenheimer, of atomic bomb fame, and his colleague George Volkoff suggested how a black hole might be born. They calculated the physical properties of a neutron core that might reside inside a massive star. They assumed it was composed of a degenerate gas and used general relativity to calculate its mass. They determined that a neutron core with a mass less than about 0.75 Suns could be gravitationally stable but a larger one will collapse without limit. (Modern estimates have raised this limit to about 3 solar masses.) John Wheeler, an astrophysicist at Princeton University, suggested the name "black hole" for such an object.

Schwartzschild's black hole is the simplest possible. We can think of it as a sphere the size of the event horizon, with a point-like mass at its center. In the 1960s and 1970s a series of brilliant theorists investigated the relativistic properties of more complex black holes. Roy Kerr worked out the effects of rotation and, with Ted Newman, clarified the effects an electric charge would have on the neighborhood. Roger Penrose described how quantum mechanics affects the structure of a hole. Stephen Hawking showed that, contrary to expectations, a hole could emit a type of radiation.

These researchers also learned that there is no limit on the mass of a black hole. A hole may have a mass comparable to a star, or a giant molecular cloud, or even a galaxy. In this respect black holes differ from white dwarfs and neutron stars.

All these concepts remained the province of a small group of relativists, without observational proof, until the early 1960s. Indeed, many astronomers didn't understand this abstract theory and were skeptical that these objects, if they existed, could ever be detected.

Then the discovery of quasars and x-ray binaries revolutionized astrophysics and placed black holes at center stage.

BRIGHTER THAN A TRILLION SUNS

The discovery of quasars has been retold many times, but it is such a good story that I cannot resist repeating it. In the 1950s radio astronomers were finding discrete sources of radio radiation all over the sky. Although most sources could be identified with particular galaxies or supernova remnants like the Crab Nebula, some lacked any connection to a visible object.

Nailing down these elusive sources was difficult, because radio telescopes of that day had poor angular resolution. They could fix the position of the center of a radio source only within a few arcminutes at best. Cyril Hazard, a radio astronomer at Cambridge University, devised a clever method to overcome this obstacle, at least for those sources that lie in the Moon's path across the sky. He would track a source with his telescope and wait for the Moon to catch up (note 10.8). From the precise time when the Moon covered the source, Hazard could fix its position in the east-west direction within a few arcseconds.

In 1962 he successfully applied his method to an unidentified point-like source, labeled 3C273 in the third Cambridge catalog of sources. Maarten Schmidt at Caltech heard about his coup and decided to examine the relevant area on the sky with the 200-inch telescope. He found a faint blue star in the error box and recorded its spectrum. It was like no other he had ever seen, a group of bright emission lines of no known element.

He struggled for weeks to understand his weird spectrum. Finally, light dawned: he was staring at the well-known Balmer lines of hydrogen, whose wavelengths were all shifted to the red by about 15 percent. If the shift were due to the expansion of space, it meant that the source was receding from Earth at 45,000 km/s. And that meant the source was at a distance of about 3 billion light-years. How could any source emit enough radiation to be detectable from such a distance?

Some astronomers could not accept that the source was really so distant. Perhaps these were merely high-speed objects in our galaxy, or perhaps some novel physics could account for the red shift. But similar objects were soon found with larger wavelength shifts. The inferred speeds meant that these objects could not remain close to our galaxy for very long. So, these "quasi-stellar" objects must indeed lie at great extragalactic distances. That implied that they emit radio energy at rates as high as 10^{35} kilowatts, which is equivalent to the total output of a trillion Suns. Moreover, the sources varied in brightness within days or even hours, which argued for very compact source sizes (note 10.9). What mechanism could account for such stupendous rates of emission from a relatively small source?

Many ideas were floated, including nuclear fusion and antimatter annihilation, but calculations eliminated all but gravitational contraction as an adequate agent. George McVittie at the University of Illinois made a crucial connection in 1964 between the observed emission of quasars and Schwartzschild's black hole. He estimated that an isolated mass of 10^8 Suns, packed into a sphere with the diameter of the orbit of Pluto, could contract to a point of infinite density and radiate 10^{35} kilowatts for 100,000 years.

Edwin Salpeter, an astrophysicist at Cornell University, added anther key concept in the same year. He showed how an object of about 10^6 solar masses could radiate energy indefinitely, by attracting the interstellar gas near it. He also estimated the luminosity of the object, in terms of the rate of mass infall.

Then in 1969 Donald Lynden-Bell of the Royal Greenwich Observatory took a bold jump and proposed that a huge black hole lies at the center of a quasar and powers its radiation. With Martin Rees, he also suggested that interstellar gas would spiral into the hole as a flat accretion disk. Friction in the disk would slow the gas and allow it to "slowly run down into the central black hole just as water runs out of a bath" (*Monthly Notices of the Royal Astronomical Society* 152, 461, 1971). Although this was an attractive model for a quasar, final proof that supermassive holes exist would have to wait until 1998 (note 10.10).

Meanwhile another candidate for a black hole surfaced.

THE X-RAY SKY

The science of x-ray astronomy was inaugurated on June 18, 1962, when Riccardo Giacconi and his colleagues at the American Science and Engineering

Corporation launched three x-ray detectors on a rocket. In a flight that lasted less than six minutes, their equipment discovered two bright x-ray sources, which they labeled Cygnus X-1 and Scorpio X-1. They would turn out to be among the brightest x-ray sources in the sky. Later, Giacconi and friends built the instruments for the Uhuru satellite, which discovered over three hundred x-ray sources.

Iosif Shklovsky, a Soviet astrophysicist with a famously fertile imagination, proposed a mechanism for the production of the x-rays. He suggested in 1967 that Scorpio X-1 was a binary star. One of its components was a neutron star, he guessed, which pulled gas off its normal companion. The gas was heated as it fell inward, and produced the observed x-rays.

Then in 1972, C. Thomas Bolton, a Canadian astronomer, determined from spectroscopic data that Cygnus X-1 is in fact a binary system. It consists of a hot blue supergiant and a dark massive companion. Bolton adopted Shklovsky's scenario and suggested, without further proof, that the dark companion was a black hole of stellar mass. The variable x-rays, he proposed, are produced by accretion of mass onto the black hole.

Most astrophysicists were skeptical of Bolton's bold hypothesis, and several of them proposed alternative models. But Kip Thorne, the black-hole theorist at Caltech, bet Stephen Hawking a subscription to his favorite magazine that Bolton was right. They would have to wait to settle the bet.

By 1976, Bolton had enough orbital data to limit the mass of the dark companion and its distance from the supergiant. He determined that the binary rotates in 5.6 days and that its components are separated by a mere 0.2 AU. (Recall that an astronomical unit, AU, equals the Earth-Sun distance, 140 million kilometers.) That meant that the companion's mass was no less than half that of the supergiant, say, between 8 and 11 solar masses. It was too massive to be a neutron star, as Shklovsky had proposed. Bolton was therefore able to exclude all proposals but the black hole hypothesis.

The first black hole had been identified, and Kip Thorne had won his bet.

Now theorists had to explain how the black hole in an x-ray binary is able to accrete mass from its companion. Gas that is dragged away from the normal star possesses angular momentum that it must shed before it can fall into the black hole. Otherwise, centrifugal forces would balance gravity and the gas would settle into stable orbits.

Several scientists adopted the view that the infalling gas would form a flat, spinning disk around the black hole. If the gas in the disk possessed some form of friction, such as viscosity, the faster-rotating rings near the hole could transfer some of their angular momentum to the slower outer rings. In this way the inner rings could lose sufficient momentum to be able to cross the event horizon and be swallowed by the hole. This was certainly plausible; in fact Lynden-Bell and Rees had proposed a similar process for the accretion disk in a quasar.

Unfortunately, the ordinary viscosity of stellar gases is far too small to account for the required friction. Some authors proposed instead that turbulence in the gas acts to couple neighboring rings. But without sufficient gas viscosity, turbulence would also be too small. Two Soviet astrophysicists, Nicolai Shakura and Rashid Sunyaev, introduced the key idea needed to solve the problem. They proposed that magnetic fields in the disk would couple different radii in the disk and enable the transfer of angular momentum.

In 1973, even before the confirmation of a black hole in Cygnus X-1, they outlined a model of an accretion disk around a black hole in a binary. They guessed that the magnetic field of the donor star would be entrained in the inflowing gas and would generate turbulence. The field would become chaotic in the disk and extremely strong near the event horizon. Turbulent dissipation would heat the gas to high temperatures near the hole. Radiation from this inner zone could be absorbed in the outer parts of the disk and drive an outflow.

The key parameters governing the accretion were the mass of the hole; the rate of mass influx; and "alpha," the ratio of the magnetic and thermal energy in the disk. They were able to set limits on alpha and therefore to make testable predictions. In particular they could show how the observed spectrum of soft x-rays could arise from the temperature gradient in the disk. Their model served as a basis for much that followed.

Stuart Shapiro and his colleagues at Cornell University built on the Shakura-Sunyaev model and introduced a fresh idea for the production of the hard x-rays that were observed. They proposed a two-temperature model in 1976 that successfully reproduced the whole x-ray spectrum from 8 to 500 keV. Like Shakura and Sunyaev they assumed that magnetic fields provide the necessary viscosity in the accretion disk. As the spinning accretion disk is sucked down into the hole, viscosity dissipates the gravitational energy.

Shapiro and friends calculated that, in this process, ions are heated far more than the electrons. Thus, ions in the inner disk reach temperatures as high as 10^{11} K while the electrons rise only to 10^9 K. The outer part of the disk remains relatively cool and emits soft x-rays. Hot electrons from the inner parts collide with the soft x-ray photons and boost their energy by the well-known inverse Compton effect (note 10.11). In this way, the hard x-rays are created. Their predicted spectrum fits the observations to a tee. The energy deposited in the ions is either swallowed by the hole or acts to drive flows.

This model and extensions of it have been adopted as the standard explanation for the radiation of the 250 known x-ray binaries. In recent years, though, we've learned that Cygnus X-1 and some binaries like it are even more bizarre than we had realized. We'll come to these features in a moment, but first let's return to some parallel developments in the quasar story.

THE CENTRAL ENGINE OF A QUASAR

After the black hole was confirmed in Cygnus X-1, theorists began seriously to consider the possibility of supermassive holes in quasars. They turned to the problem of explaining the enormous radio emission from quasars and naturally adopted some of the same concepts that were successful in x-ray binaries.

However, conditions are far more extreme in the vicinity of a mass of 10^6 to 10^8 solar masses. Space-time is warped and dragged near a rotating hole, and particle velocities can approach the speed of light. Roger Blandford, originally at Cambridge University, was one of the first to design a model for these extreme objects. In 1977 he and colleague R. L. Znajek described how a supermassive black hole could convert its spin energy into radiation.

They pictured a differentially rotating accretion disk that generates a magnetic field (fig. 10.3). The field threads through the event horizon of the hole (labeled H in the figure) and couples the disk to the rotating hole. If certain conditions are met, a strong electric field is established outside the event horizon. At random points, the electric field can become strong enough to sponta-

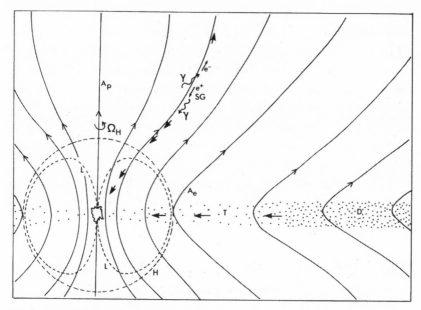

Fig. 10.3. A model of a supermassive black hole at the center of a galaxy. The dashed circle (H) represents the event horizon, the "surface" of the hole. An accretion disk (dotted slab: D, T) falls into the hole. Electron-positron pairs (e⁻, p⁺) are created at locations (like point Y) where an intense electric field breaks down. Electrons escape at near the speed of light and radiate over the whole electromagnetic spectrum.

neously break down into pairs of electrons and positrons. These particles are accelerated by the electric field to relativistic energy. Positrons flow into the hole, but electrons escape. As they flow along the curved field lines, they are forced to radiate gamma rays. These, in turn, can create secondary electron-positron pairs, and the process continues in a cascade. Eventually the electrons lose sufficient energy to radiate softer x-rays.

When Blandford and Znajek applied their model to explain a quasar, they learned that the magnetic field in the inner disk could be quite modest. To produce a radio-wavelength power of 10^{35} kilowatts, for instance, a hole with a mass of about 10^9 solar masses would require a minimum magnetic field of only 100 gauss.

During the following decades many models were advanced to explain the radio waves, gamma rays, and x-rays emitted by quasars. In 2001 Blandford summarized some of the main themes. He pointed out that two sources are able to power a quasar: the accretion disk or the spin of the black hole. The magnetic torque that transports angular momentum radially outward in a disk also transports energy. This energy can be radiated from the disk or emerge as a hydromagnetic wave that heats a corona above the disk. The corona may, in turn, drive a wind.

Alternatively, the spin of the hole may be coupled to the disk by means of a large-scale magnetic field. Again, there are several ways in which the spin energy can be dissipated. Perhaps the most interesting, from the standpoint of magnetic fields, is the ejection of a jet from one or both poles of a rotating hole. The double radio source Cygnus A (fig. 10.4) contains a prime example of such a jet.

AN ENORMOUS BALLOON

Walter Baade and Rudolf Minkowski photographed Cygnus A in 1952 with the 200-inch Palomar telescope. They discovered two luminous clouds about a minute of arc apart, which led them to think they had detected two galaxies in collision. In 1984, Richard Perley and his colleagues obtained the sharp image you see in fig. 10.4 with the Very Large Array in New Mexico. It has a dumbbell shape, with two lobes connected by a thin jet. The lobes are filled with swirling filamentary material. What is not obvious is the gigantic scale of this object. At a distance of 700 million light-years, it stretches 500,000 light-years across the sky. But it pales in comparison to 3C236, another double source, which measures 20 million light-years in size.

Researchers in the field agree that a supermassive black hole very likely resides at the bright point in the center of each of these objects. They think the hole rotates rapidly and ejects collimated jets of plasma from the ends of its rotational axis. The jets are observed in radio waves, visible light, x-rays, and gamma rays. At least some of the emission is synchrotron radiation, which argues for the presence of magnetic fields and relativistic electrons. Near the hole, blobs in the jets reach speeds as high as 98 percent of the speed of light. Their initial composition is unknown but may consist of electrons, positrons, and protons. What is certain is that gamma rays with photon energy of a trillion electron volts are generated near the hole. That speaks of electrons with comparable energy.

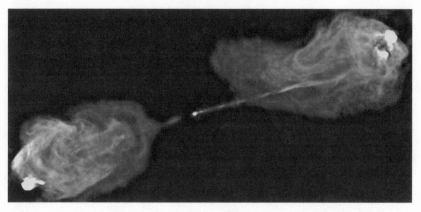

Fig. 10.4. The giant double radio source Cygnus A. This image was obtained with the Very Large Array at a wavelength of 6 cm. The bright central spot contains the black hole, the source of the jet.

How do these wonderful machines accelerate jets to such speeds? How do the jets remain focused over tens of thousands of light-years? Theorists have models but no final answers.

Their basic idea is that a rotating magnetized accretion disk surrounds the black hole. The disk winds up the field into a conical shape at each rotational pole. Somehow, gravitational and rotational energy is converted to the energy and momentum of supersonic plasma jets. Somehow, the twist in the field focuses the jets.

Roger Blandford and Donald Payne presented an influential model along these lines in 1982 (see chapter 8). In their picture a spinning accretion disk throws off gas along the inclined lines of force that thread through it. Close to the disk's surface, the pressure in a hot corona helps to launch the gas, and centrifugal forces accelerate it. At large distances from the disk, the field twists around the rotation axis and squeezes the outflow into two antiparallel jets. So it is the twist of the field that preserves the jets' focus. They calculated that a field strength of 10^2 to 10^4 gauss would be sufficient.

In their model, the jet has a core of plasma that carries off most of the energy, while the magnetic field in the walls of the jet carries off most of the angular momentum. The remarkable feature of their model is its prediction of jet speeds of a fraction of the speed of light, as observed in Cygnus A.

Models of black hole jets have proliferated almost without limit since

Blandford and Payne published their pioneering work. The models differ in how they treat three main questions: How is a jet accelerated near the hole? How is the jet collimated close to the hole? And how is the collimation maintained over distances of tens of thousands of light-years?

The subject is in a state of wild ferment, with no consensus in view. But theorists have recognized that magnetic fields are essential components of any viable model. A remarkable observation of a jet in the first quasar to be discovered, 3C273, has reinforced this conclusion.

In 2005, Robert Zavala of the U.S. Naval Observatory and Gregory Taylor of the National Radio Astronomy Observatory obtained an extraordinary radio image of this quasar with the Very Long Baseline Array (VLBA). This chain of ten identical radio telescopes stretches from New Hampshire to Hawaii. It resolves structures in distant quasars as small as a few milliarcseconds.

Zavala and Taylor wanted to test Roger Blandford's 1993 prediction that the helical magnetic field is less tightly wound in the core of a jet than at its edges. They were able to resolve the cross-section of the jet with the VLBA and discovered that the reverse is true.

Back to the drawing board!

SMALL BUT BEAUTIFUL

It doesn't take a supermassive black hole to generate a spectacular jet. Even a modest hole of a few stellar masses can turn the trick. Cygnus X-1, that reliable astrophysical laboratory, provides a perfect example. In 2001, A. M. Stirling and colleagues from several radio observatories reported the discovery of a thin synchrotron-emitting jet emerging from Cygnus X-1. More recent observations show that the jet is 30 AU long and moves at one-third the speed of light. It varies in length from 30 to 140 AU.

Then to top this discovery, Elena Gallo and her international team discovered a luminous bubble 16 light-years in diameter at the end of the Cygnus jet. Apparently the jet is inflating the bubble in much the same way the jet of Cygnus A inflates its double lobes. The team estimates that the power output of the jet (10^{26} to 10^{27} kilowatts) could equal the total x-ray emission of the binary. They suggest that the bulk of the power liberated by accretion escapes as dark outflows rather than emission near the hole. This result has given theorists much food for thought.

In recent years researchers have realized that jets appear in many different astrophysical situations where a magnetic field surrounds a compact mass. They are detected in young stellar objects, like T Tauri stars; x-ray binaries, like Cygnus X-1; pulsars; and in galaxies with active nuclei, like quasars. It may not be possible to find a universal model that could apply to all these objects, but it is clear that some of the same physical principles are involved.

THE GALAXIES

IN THE MID-EIGHTEENTH CENTURY, astronomers like Charles Messier were finding faint blurry "nebulae" all over the sky. Messier considered them a nuisance, because his prime interest was finding new comets and the nebulas distracted him. He decided to compile a catalog of them so as to avoid them in the future. On the night of October 13, 1773, he spied yet another, in the constellation Canes Venatici. He noted its position and labeled it M51 in his catalog. We would recognize it two centuries later as the Whirlpool Galaxy (fig. 11.1).

Sir William Herschel, the eminent British astronomer, received a copy of Messier's catalog from a friend on the occasion of his election to the Royal Society. The catalog awakened his interest in nebulas. He had read Immanuel Kant's speculation that the nebulas were "island universes," great collections of stars separated from us by immense distances. By counting the density of stars in the sky, Herschel had already determined that the Milky Way was a great flat disk of stars and could be one of Kant's island universes. He decided he would observe all one hundred nebulas on Messier's list with his large telescopes and see what he could discover. Caroline, his sister and faithful assistant, would record his observations.

On May 12, 1787, he observed M51 through hazy skies with his 24-inch telescope. He noticed "a very uncommon object, nebulosity in the center with a nucleus surrounded by detached nebulosity in the form of a circle, of unequal brightness in three or four places, forming altogether a most curious object" (J. L. E. Dreyer, *The Scientific Papers of Sir William Herschel*, Royal Society and Royal Astronomical Society, London, 1912). This was as close as

Herschel got to see spiral arms in the misty English air. But he went on to find and catalog over 2,500 nebulas.

William Parsons, third Earl of Rosse, was luckier. Parsons graduated from Oxford in 1822 with first-class honors in mathematics, but his real love was astronomy. When he succeeded to his title and the fortune that came with it, he decided he would build a telescope with a mirror larger than Herschel's 48-inch. Moreover, he would cast the mirror himself out of speculum, a shiny metal alloy. To learn the tricks of the trade, he first cast a 36-inch mirror. No sooner had he finished than he boldly decided to attempt a 72-inch. He had to struggle, but he did succeed, and the completed telescope was judged to be excellent by the experts.

One of Parsons's greatest joys was observing with his fine telescope. One night in 1845 he turned it to M51. He could see its spiral arms clearly. Indeed, he was the first ever to see such arms. Struck by the nebula's beauty, he made a remarkably precise drawing of it. He called it the Question Mark Galaxy, and others referred to it as Rosse's Galaxy, but we know it as the Whirlpool. He suspected that it could be a great rotating assembly of stars. He was correct, of course.

In the century that followed, M51 and spirals like them were studied in great detail. We learned that the arms contain clusters of the youngest stars, as well as most of the dust and hydrogen gas. The disk in which the arms are embedded contains older stars, like our Sun. As telescopes and detectors steadily improved, we learned more and more about the structure of spirals. Finally, we were treated to the spectacular image from the Hubble Space Telescope that we see in figure 11.1.

Radio astronomers have given us similar images of the Whirlpool, not quite as sharp as those from the Hubble, but showing something different: the well-ordered magnetic field that follows the shape of the spiral arms. In the past two decades, observations at centimeter and millimeter wavelengths have given us detailed information on the fields in the Whirlpool and other spirals, including the Milky Way. Theorists are working on models to explain the new features and to understand how the fields originated.

In this chapter we'll recall some of what they've learned. We'll begin with some background on the structure of spiral galaxies and the basic theory of the origin of the arms. Then we'll meet some of the radio astronomers and find out what they've learned about the magnetic fields. Finally, we'll see what the experts have to say about the origin of galactic fields.

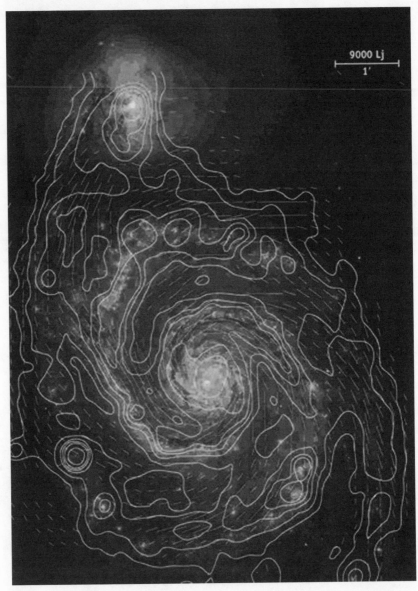

Fig. 11.1. The Hubble Space Telescope obtained this image of the Whirlpool Galaxy (M51). Luminous gas clouds and stars outline the spiral arms (grayscale). The superposed contours show the intensity of synchrotron emission at a wavelength of 6 cm, and the short lines show the intensity and direction of the ordered magnetic field. The observations were made at the Very Large Array and the Effelsberg 100-meter telescope. The bar indicates a length of 9,000 light-years.

A PRIMER ON GALAXIES

Edwin Hubble, the man we usually associate with the expansion of the universe, returned from France in 1919 after serving as an infantry major in the Great War. He was uncertain about a choice of career. He had acquired degrees in physics, Roman and English law, and astronomy. He had practiced law before the war and disliked it. George Ellery Hale, the director of the Mount Wilson Observatory, had offered him a job as junior astronomer in 1913, but he had enlisted instead. When he returned he finally decided that he would pick up Hale's offer.

The rest is history, as they say. Hubble embarked on a systematic study of the recession velocities of galaxies, using the 100-inch telescope to great advantage. His work led eventually to one of the great revolutions in science. His raw material was photographs of galaxies, hundreds of them. He soon noticed smooth variations in shape and decided to arrange the galaxies in some sort of order. He hoped to discover an evolutionary sequence.

Hubble sorted galaxies into three main categories: elliptical; spiral with a central bar; and spiral without a bar. The elliptical galaxies were graded by their amount of flattening; the spirals by the tightness of their winding and the size of their central bulges.

Hubble's categories for spirals have been somewhat revised and relabeled. Grand design galaxies like M51 have two well-defined arms and a small central bulge. "Flocculent" galaxies are at the other extreme, with many incomplete and weakly defined arms. "Lenticular" galaxies are a transition type between elliptical and spiral. They have large central bulges and no arms.

Some spiral galaxies, like the Andromeda Nebula (M31) are tilted enough to show that the arms lie in a flat disk. "The Pencil" in the constellation Coma Berenices (Berenice's Hair) is turned to face us edge-on. It demonstrates that a spiral galaxy is essentially a flat disk of gas and stars, with a central bulge and a dim spherical halo.

The Milky Way Galaxy

Galaxies range in size from giants, like the Milky Way and its neighbor M31, to so-called dwarf galaxies like the Small Magellanic Cloud. The Milky Way's disk has a diameter of about 30 kiloparsecs, a thickness of about 0.4

kiloparsec, and contains between 200 and 400 billion stars. (One parsec equals 3.26 light years; see note 11.1.) Surrounding its flat disk is a halo of million-degree plasma and a spherical array of globular clusters, each of which contains thousands of very old stars. Our solar system is located about halfway from the center and revolves about the center in around 240 million years.

We now know that our galaxy is a spiral with five arms, but it took a major effort to locate them because of our unfavorable location in the plane of the disk. The breakthrough came shortly after the end of World War II.

Jan Oort, a Dutch astronomer, learned that an American engineer, Karl Jansky, had discovered cosmic radio waves. Another American, Grote Reber, had published maps of such radio sources as the center of the Milky Way, at a wavelength of 2 meters. Oort realized that radio waves could penetrate the dense dust clouds that had frustrated optical astronomers and open up the galaxy to further exploration.

He was especially interested in finding the radio equivalent of a spectral line, which would allow him to measure the Doppler velocities of interstellar hydrogen clouds. In turn, that might enable him to determine the positions of the spiral arms. So he put his graduate student Hendrik van de Hulst to work searching for a spectral line of atomic hydrogen in the radio spectrum.

It didn't take him very long. Van de Hulst predicted the existence of a line at 21 cm in 1944 (note 11.2). After the war, the Dutch tried unsuccessfully for several years to detect the line. Later, Harvard professor Edward Purcell and his student Harold Ewen built a special radio receiver and confirmed the existence of the spectral line. After Oort learned of their novel technique, he borrowed some German radar equipment and was able to confirm their detection within a few months. In 1952, Purcell shared the Nobel Prize in Physics (with Felix Bloch), partly in recognition of the discovery.

Oort went on to measure the velocities and positions of hydrogen clouds in the galactic plane. His method was simple. He assumed that the clouds were located mainly in the arms; that they revolve in circles around the center; and that the smaller the circle is, the faster the clouds revolve.

In each direction in the galactic plane, he detected several clouds with different Doppler velocities. He reasoned that the cloud with the maximum velocity must lie closest to the center, and from the geometry he could calculate its distance from the center. From these data he could construct a map of the differential rotation of the disk. Then he could go back and, from the speed of each individual cloud, place it along its line of sight. Finally, the map of the clouds revealed the shape of the arms. He discovered four arms and traces of a fifth.

Later, Frank Kerr in Australia extended Oort's map to the southern parts of the Milky Way, and many further refinements were made. We now know the galaxy has five arms, named in order from the center: the Norma; Crux; Sagittarius; Orion; and Perseus arms. We live in the Orion arm.

Although the structure of the galaxy's arms has been sorted out, its center is still something of a mystery, as it is shrouded in dense clouds of dust. But in 2002 observers at the European Southern Observatory completed a decade of measuring the orbit of a star that revolves within 17 light-*hours* of the center. The star's short period of fifteen years indicates the presence of a supermassive black hole of 2.6 million solar masses at the center.

Vera Rubin and Kent Ford at the Carnegie Institute of Washington discovered something even more surprising in 1970. They learned that, beyond a certain radius in M31, the orbital speeds of ionized hydrogen clouds (in km/s) hardly varied along an outward radius. They had expected that the speeds would decrease toward the outer edge of the galaxy, as it does for the planets in our solar system.

Later observations showed that the same effect prevails in many galaxies, including the Milky Way. These results implied that more mass exists in galaxies than is associated with the visible material. Some unknown form of dark matter is present in all galaxies that alters their differential rotation. Astronomers now know that dark matter constitutes more than 20 percent of the total mass of the universe, but nobody knows for sure just what it is.

Spiral Arms

Several theories have been proposed to explain the origin of spiral arms, but none is completely satisfactory. The theory conceived by C. C. Lin and Frank Shu in 1964 goes a long way toward explaining the grand design spirals but breaks down for the flocculent variety. Other theories perform somewhat better for those. It may be that a combination of models works best.

The basic issue in spirals is the "winding problem." As Rubin and Ford discovered, the speed of orbiting gas clouds (measured in km/s) hardly varies from a galaxy's center to its edge. Clouds in larger orbits, accordingly, take longer to complete a revolution than those in smaller orbits. So an arm composed of a *fixed* group of clouds would wind up tightly within a few rotations. But most spiral galaxies have completed several tens of rotations during their lifetimes and still are observed to possess discrete arms. How can this be?

One possibility is that the arms we see are actually transient; they are in fact winding up and will blur together after a few more rotations. From this point of view, the arms in the flocculent spirals could either be decaying or forming. But observations in grand design spirals suggest otherwise. Astronomers observe that their arms contain both short-lived and long-lived stars. That argues that the arms are long-lived and therefore must be a *pattern*, rather than a material object.

Lin and Shu adopted this view. They were professors at the Massachusetts Institute of Technology in the 1960s and agreed to collaborate on this fundamental problem. Shu, you may remember from chapter 8, contributed important ideas on the birth of stars. Lin is a fluid dynamicist who has spent most of his career studying galactic structure.

They proposed that spiral arms are "density waves" that rotate around the center of a galaxy as a permanent pattern. The stars and clouds that we see in an arm are constantly changing, they said, constantly moving through the arm. But the pattern remains the same, presumably for billions of years.

Sound waves are also density waves, and they offer some help in picturing a spiral arm. Unlike a bullet, a sound wave is not composed of a fixed group of molecules. Instead, different molecules along the wave's path oscillate a short distance around their home positions as the wave arrives and passes. The air compresses as the wave arrives and then decompresses.

Something similar occurs in a spiral arm. Stars catch up with the back of an arm, slow down slightly, and ride with the arm for a while and then leave from the front of the arm. This small oscillation in the orbiting speed of a star requires moving only a short distance from its normal orbit. So the density of stars within the arm remains roughly constant, but the stars within it are constantly changing.

That's the essence of the idea, but we need to delve a bit deeper. A star in the disk experiences the gravitational pull of all the other stars in the disk as well as those in the massive central bulge. As a result a star's orbit is *not* a closed curve, like those of the planets in the solar system (note 11.3).

As a good approximation, however, its orbit is an ellipse that precesses (revolves) around the center at a constant rate (fig. 11.2). Keep in mind that the star revolves in its orbit at the same time the orbit itself precesses around the center. In the simple model we're discussing, all the nested ellipses precess at the same rate, but the inner ellipses are slightly tilted with respect to the outer ellipses, so that they nearly overlap for some distance. *This overlapping portion forms a part of a spiral arm.* Depending on the relative tilts, two or three or more arms can form in this way. The pattern of arms revolves at the common precession rate of all the orbits.

This still isn't the whole story, because we've neglected the enhanced gravity of the arm itself, which acts on stars that approach and leave it. But this will do for a start.

It's perhaps easier to understand how a spiral pattern might persist than to grasp how the pattern is set up in the first place. Lin and Shu didn't really answer that question initially. They did show that a thin differentially rotating disk of stars (or gas) would be unstable if it were perturbed slightly. Rotating ripples would develop spontaneously and some would grow faster than others. Such disturbances are seen in the rings of Saturn, for example. Lin and Shu interpreted the spiral arms we see as the fastest-growing instabilities.

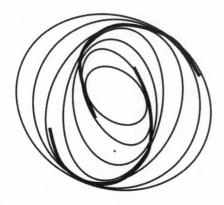

Fig. 11.2. A cartoon to explain the origin of spiral arms as a density wave. Stars and clouds orbit the galactic center in ellipses. The combination of gravitational forces causes each ellipse to tilt slightly and to precess around the center at a constant angular speed. Where the ellipses nearly touch, stars bunch up in a pattern of spiral arms.

They showed that two arms are the most likely result and proved later that the arms could be permanent.

Other authors have proposed that some discrete event triggers the formation of spiral arms. The near collision of two galaxies, for instance, could generate tidal forces that start the process. Or possibly a sequence of supernovas within the disk could excite the arms. The issue is still open.

The Milky Way's Magnetic Field

As you might expect, the Milky Way was the first galaxy in which a magnetic field was discovered. In 1949, William Hiltner at the Yerkes Observatory and John Hall at the Lowell Observatory independently discovered that starlight in the galaxy is polarized by a few percent (note 11.4). Hiltner was puzzled at first, because a thermal source like a star shouldn't emit polarized light, but he knew that the spaces between the stars are filled with clouds of dust grains.

Not a lot was known about them except that they absorb light efficiently. Great dark clouds of them are visible even to the naked eye. The Coal Sack in the southern constellation Crux is a good example. So Hiltner proposed that elongated dust grains might absorb starlight in such a way as to produce the observed polarization. That would require a force, perhaps a magnetic field, to align the grains.

Leverett Davis Jr. and Jesse Greenstein, astronomers at Caltech, offered a more complete explanation in 1950. They proposed that elongated grains spin at high speed around a uniform interstellar magnetic field (note 11.5). The component of starlight that is polarized perpendicular to the field would be absorbed preferentially, which would lead to a net polarization of the transmitted light. The astronomers estimated that a field of only a few microgauss could explain the observations. Later, Hiltner observed the same effect in M31 and discovered that the field was aligned along the spiral arms.

By 1970, 7,000 stars in the galaxy had been measured for interstellar polarization. Australian astronomers Donald Mathewson and Victor Ford collected all the data in a remarkable map. It showed that magnetic fields are mainly aligned parallel to the plane of the disk but with marked irregularities. The real surprise was the great loop of field above the disk that extends

120 degrees across the sky. It seems to lie nearby, between 100 and 200 pc of Earth.

This method of detecting magnetic fields is not as sensitive as those we will consider in a moment, but it is still useful at submillimeter wavelengths. For instance, David Chuss and his colleagues at the NASA Goddard Space Flight Center have recently measured the polarized emission of dust grains only 50 pc from the center of the galaxy, at a wavelength of 0.35 mm. They found fields that could be as strong as a few milligauss in dense clouds.

Probing the Milky Way with Pulsars

As we learned in chapter 10, pulsars emit powerful beams of polarized radiation at radio wavelengths. As a pulsar's beam passes through ionized hydrogen clouds, its plane of polarization rotates by an amount that is proportional to the square of the wavelength and to a quantity called the rotation measure, RM. The RM is the product of the average field strength and the total number of electrons along the line of sight. This effect is called Faraday rotation, after its discoverer (fig. 11.3). The *sign* of the RM depends on the direction of the field. Therefore, from measurements of the RM along a fixed line of sight, at several wavelengths, an astronomer can estimate both the average strength and the direction of the field along that line of sight. And by combining the RM on different lines of sight to different pulsars, she can construct a model of the field in two or even three dimensions (note 11.6).

This may sound straightforward, but it's not, and different researchers have come to different conclusions about the shapes and directions of the fields in our Milky Way. Some facts are agreed to. Pulsar rotation measures in the 1970s showed that the field in the local (Orion) arm turns *clockwise* (CW), as seen from north of the galactic plane. Then in 1980, Canadian astronomers Martine Simard-Normandin and Phillip Kronberg analyzed the rotation measures of some five hundred extragalactic radio sources. They discovered a *counterclockwise* (CCW) direction in the neighboring Sagittarius arm. Beyond those two well-established results, there is controversy.

Canadian astronomer Jacques Vallée used his collection of 350 pulsar rotation measures to derive the field direction in the disk without identifying individual arms. He concluded in 2005 that the field is CW overall from

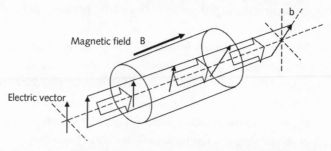

Fig. 11.3. A cartoon to illustrate Faraday rotation. We see a plane-polarized wave, in which the electric vector vibrates in the vertical direction as the wave progresses. As the wave traverses a column of plasma in which a parallel magnetic field (B) exists, the plane of polarization rotates by the angle b. The size of the angle depends on B, the number of electrons along the column, and the wavelength. In general, waves with short wavelengths rotate less.

center to edge, except for a band between 4 and 6 kpc from the center. (Recall that the Sun lies about 8.5 kpc from the center.)

Jin Lin Han, at the Chinese Academy of Sciences, has presented the most controversial results. Over the past decade he assembled Faraday rotation measures for 374 pulsars distributed nearly uniformly over the sky and derived from them the magnetic fields in several arms of the Milky Way. Although he admitted some flexibility in modeling the data, he concluded in 2006 that the direction of the field is CCW in all five arms except the Orion arm, in which we are located, and *reverses in every lane between arms.* His assertion is the more striking because such reversals have rarely been seen in other galaxies.

The latest study in one quadrant of the galaxy combines 149 extragalactic and 120 pulsar rotation measures. Jo-Anne Catherine Brown at the University of Calgary and her international team concluded in 2007 that the field is CW there (fig. 11.4). Notice, however, that there is some evidence for CCW field in the Perseus and Norma arms. But on the whole they don't agree with Han that many reversals are needed to model the data.

From all this you can see that a large-scale field in the galactic disk is still open to debate. The field in the halo is also controversial because of the difficulty of separating intense small-scale structure from the overall pattern. Most observers agree that the field on each side of the disk is a torus that is centered on the bulge but disagree on whether the tori have the same or op-

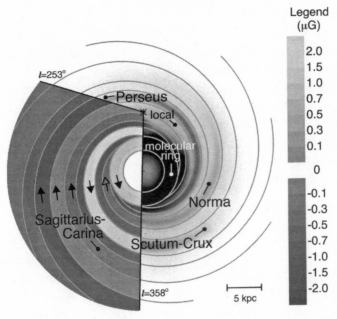

Legend
(μG)

2.0
1.5
1.0
0.7
0.5
0.3
0.1

0

-0.1
-0.3
-0.5
-0.7
-1.0
-1.5
-2.0

l=253°

Perseus

local

molecular
ring

Sagittarius-
Carina

Norma

Scutum-Crux

l=358°

5 kpc

Fig. 11.4. A recent map of the direction of magnetic fields in the arms of the Milky Way. Two galactic longitudes *l* are shown. Shading represents the intensity of the magnetic field in μG.

posite field directions. Han concluded from his analysis of rotation data that the tori are opposite; P. Frick and colleagues decided that they are parallel.

The shape of the field at the galaxy's center is still a puzzle. For the past two decades, radio astronomers have been exploring huge synchrotron-emitting filaments, in which well-ordered fields are aligned predominantly *perpendicular* to the galactic plane. Recent observations of cold dusty clouds reveal, though, that a strong field also exists *parallel* to the plane.

Giles Novak and his colleagues at Northwestern University, in Evanston, Illinois, made the discovery in 2003. They observed molecular clouds within 500 parsecs of the center with a 2-meter telescope at the South Polar Station in Antarctica. They recorded the polarization of the light at a wavelength of 0.450 mm, which reveals the magnetic alignment of the dust. Their map confirmed that, indeed, some strong fields are aligned parallel to the galactic plane. The field is evidently poloidal in some parts and toroidal in others.

To sort out this tangle, David Chuss and a team of radio astronomers traveled to Mauna Kea in Hawaii to use the Caltech Submillimeter Observatory.

They observed molecular clouds within a mere 50 parsecs of the galactic center, and at a wavelength of 0.350 mm. They discovered that the field in *dense* clouds lies parallel to the galactic plane and the field in *tenuous* clouds runs perpendicular to the plane.

The team offered two possible explanations for their results. Perhaps the weight of the gas in a dense cloud distorts an originally poloidal field into a toroidal field. Or perhaps the winds from supernovas distort an originally toroidal field in the tenuous clouds. Clearly, more work will be needed to choose between these alternatives, or to fashion some new ones.

MAGNETIC FIELDS IN OTHER GALAXIES

Pulsars are generally too faint to detect in distant galaxies, so radio astronomers rely on a galaxy's intrinsic synchrotron emission to map its magnetic field. Cosmic ray electrons trapped in the galactic field are the sources of this radiation.

From the total intensity of the synchrotron emission, one can determine the strength of the field; from the direction of the polarization, one obtains the direction of the field projected on the plane of the sky. Rainer Beck, senior astronomer at the Max Planck Institute for Radio Astronomy in Bonn, Germany, has observed dozens of galaxies in this way. Since 1979 he's used the 100-meter steerable dish at Effelsberg. Figure 11.1 is an example of the excellent resolution that he obtained with this telescope and the Very Large Array in New Mexico.

Lately he's been arguing strongly for the construction of the Square Kilometer Array, which would have a hundred times the sensitivity of the Effelsberg telescope and angular resolution better than the Hubble Space Telescope. It is the heart's desire of a large international group of radio astronomers.

Beck has reviewed the status of galactic magnetic fields several times during the past decade. Here are some highlights from his observations of synchrotron emission.

Magnetic arms are observed in all varieties of spiral galaxies, including the flocculents, which have diffuse arms, and the irregular galaxies, which have none. Somehow the magnetic structure seems to predate the formation of optical arms.

Spiral galaxies have average field strengths between 6 and 15 microgauss. In grand design galaxies like M51 the average field strength is about 15 microgauss. The largest fields (as high as 50 microgauss) are found in the arms of "starburst" galaxies, where stars are being born at a high rate. In those regions the fields are both strong and highly turbulent.

The arms of spiral galaxies contain both aligned (that is, resolved) fields and (unresolved) tangled magnetic fields. The aligned fields in the arms are generally weaker (1 to 5 microgauss) than the tangled field, but the opposite is true in the interarm regions. M51 is exceptional, with ordered fields of 15 microgauss at the inner edges of the inner arms.

In the Milky Way the field strength is about 6 microgauss near the Sun and increases to 20 to 50 microgauss near the center. The field lies along the arms, parallel to the plane of the disk. Strange twisted filaments lie perpendicular to the disk, with field strength as large as a milligauss.

As we can see in figure 11.1, the resolved field follows the spiral pattern of the optical arms. These spirals are usually well described as logarithmic spirals, in which the angle between the tangent at any point and the radius from the center form a constant angle, the so-called pitch angle. Pitch angles vary among galaxies from about 10 to 40 degrees. The smaller the angle, the more tightly wound is the galaxy. As we shall see, the pitch angle is a critical test for any model of the origin of the field.

Galaxies that lie edge-on to us yield information on the magnetic fields in their halos. The aligned field lies parallel to the disk and decreases in strength away from the disk, falling by 50 percent within 1 or 2 kiloparsecs. Researchers are debating whether the halo fields are toroidal and, if so, whether the direction of the toroid flips from one side of the disk to the other. Theorists can predict either outcome.

So far I've described results from observations of the synchrotron radiation. Another technique for measuring galactic magnetic fields depends on the Zeeman effect, the splitting and polarization of spectral lines. It yields direct measures of the field's strength and direction, but its application is limited to protostellar cores, where molecules like carbon monoxide and hydroxide abound. I referred to some measurements by Richard Crutcher in chapter 8. In cores, the fields reach several hundred microgauss and are remarkably straight.

You may have noticed that I haven't mentioned magnetic fields in elliptical galaxies. In fact very little is known about them, because the usual signatures are missing. Elliptical galaxies lack dust; therefore, polarized dust emission is absent. Supernovas are infrequent, so relativistic electrons, which are necessary for polarized synchrotron emission, are rare. And cold atomic hydrogen is also rare, which eliminates detection by the Zeeman effect. Some information is dribbling out from studies of their x-ray emission, but we'll pass that by.

CHICKEN OR EGG?

Two theories for the origin of galactic fields have vied for attention for the past fifty years. One theory views the fields as "primordial": they originated in the formation of the universe and were trapped and amplified during the formation of the individual galaxies. The opposite point of view sees the fields continually regenerated by some type of dynamo in the galactic disk. In that theory, the "seed" for the dynamo might have been primordial.

The question whether either theory is valid has been debated since the 1960s. This may seem surprising, considering how much observations improved over the past half century. One might think we'd have sufficient data by now to validate or disprove at least one of the theories. And indeed, the dynamo theory appears to have won out, although not without some lingering reservations.

Enrico Fermi launched the primordial theory as a conjecture in 1949, shortly after Hiltner's discovery of galactic fields. Fermi was a charismatic nuclear physicist who contributed to the development of quantum physics, statistical mechanics, and the atomic bomb. He was awarded the Nobel Prize in Physics in 1938 and had a subatomic unit of length named after him.

In a paper on the origin and acceleration of cosmic rays, Fermi showed how charged particles could gain energy if they bounced between magnetized clouds in a galaxy. Fermi had no idea how the fields originated, but he was aware that once they were established, they would decay very slowly. Therefore, he suggested, the fields we observe today could be fossils.

Fred Hoyle, his equal in stature, disagreed. He pointed out the "wrapping problem." If the galactic field were initially uniform and straight, the differential rotation of the disk would wrap it into a spiral within one or two rotations. After the fifty or so rotations a galaxy has rotated since its formation, the spiral would be wound so tightly that adjacent lines of force would alternate in sign every 100 parsecs along a radius. That would wipe out any signal of interstellar polarization, contrary to Hiltner's observations.

Eugene Parker, whom we have met repeatedly in earlier chapters, added another nail to the coffin. He countered Fermi's conjecture by pointing out that a primordial field couldn't survive for the life of a galaxy; it would escape through the short dimension of the disk by ambipolar diffusion in just a few hundred million years.

Of course, Parker had his own explanation for the ordered galactic fields. In 1955 he had discovered a way to construct an astrophysical dynamo. The essential elements were differential rotation in a conducting gas, and cyclonic turbulence, as we outlined in chapter 3. A galaxy might satisfy both of these requirements. So in 1971 Parker in the United States (and independently S. I. Vainshtein and Alexander Ruzmaikin in Russia) sketched a theory of a galactic dynamo in a thin disk.

Other researchers, such as Max Steenbeck, Fritz Krause, and Karl-Heinz Rädler, picked up the dynamo concept in the 1960s and developed it into a full-fledged theory, called mean-field magnetohydrodynamics (see chapter 2). Later they applied it to rotating galaxies. The basic scenario runs as follows.

The differential rotation of the disk stretches primordial lines of force and winds them up around the center (the so-called omega effect). Simultaneously the rotation induces helical motions in the turbulent ionized gas. Coriolis forces ensure that the helices are cyclonic, that is, they have a preferred sense of twist. The helices twist the toroidal field lines into small-scale loops (the alpha effect) that all have the same sense of twist. The loops are, as a result, able to reconnect and merge to form a large-scale poloidal component. In a steady state, the rotation generates the poloidal field from the toroidal field, and vice versa. Round and round it goes.

There is a problem, though: a frictionless dynamo would work too well, like the sorcerer's apprentice. It would generate a field continuously and at a constantly increasing rate. Without some way to dissipate or eject the field lines, the field strength would grow without limit. One way out of this trap is to assume that as the field grows, so does the turbulence in the ionized gas. Small-scale turbulence would allow tangled field lines to reconnect and cancel. The energy contained in the field would be converted to heat and radiation. Ideally, this process would come to an equilibrium, in which the turbulence and the field would contain equal amounts of energy in each cubic meter, a condition called equipartition. The average field strength would then level off and remain roughly constant.

Alternatively, the field might escape out the top and bottom of the disk at a rate that would balance the rate of generation. But the devil is in the details. Theorists would have to demonstrate how either or both of these mechanisms might work in practice.

In the mid-1980s, dynamo models by T. Sawa and M. Fujimoto at Nagoya University and by Alexander Ruzmaikin at Moscow University reproduced both the one-armed and two-armed spirals observed in galaxies, and the dynamo mechanism was accepted as a viable explanation. But there were some limitations and a few definite worries. Robert Rosner and Edward DeLuca, students of Parker, pointed out some in 1989.

First, the theoretical models were kinematic, meaning that the motions were all specified in advance. Second, the models were linear, meaning they did not include the back-reaction of the field on the motions. As a result, the models couldn't predict the final strength of the fields. Third, no adequate theory existed for the turbulent diffusion that controls the dissipation of fields. Finally, the models only applied to the ionized gas, without considering the effect of the neutrals on the motions. Nevertheless, such dynamos could produce realistic-looking spiral fields; figure 11.5 is a recent example.

In the following decades, several of these limitations were overcome. A number of researchers constructed nonlinear models of varying complexity that not only resembled real galaxies but also predicted reasonable field strengths. In one model Axel Brandenberg, a Danish researcher, showed how the field strength would first grow exponentially and then saturate at about twice the equipartition value in about 10 billion years.

The weak point in all these models was the calculation of the alpha effect, the twisting of toroidal field in the disk to poloidal field that keeps the dynamo running. Theorists had to include the effect with rather ad hoc assumptions that would simulate cyclonic turbulence. In 1993 Katia Ferrière, a French theorist, offered an idea that was based on observations rather than supposi-

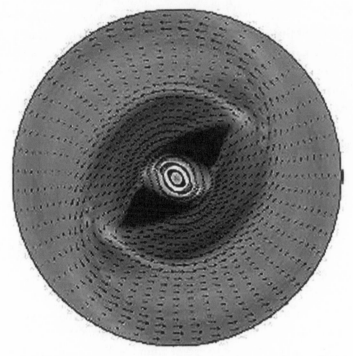

Fig. 11.5. A simulation of spiral arms in a barred spiral, with a kinematic dynamo. Shades of gray show gas density, with lighter shades corresponding to higher values. In this model the turbulence was enhanced by a factor of 2 to 6 in the dust lanes.

tions. She proposed that supernova explosions in the disk were primarily responsible for the cyclonic turbulence.

She pointed out that clusters of supernovas were observed to produce so-called superbubbles in the interstellar medium. Then she showed how these expanding bubbles stir the gas, and she derived formulas to describe how these motions create the alpha effect. With her colleague, Dieter Schmitt, she generated a series of models to test the idea, with encouraging results. A nonlinear, time-dependent model generates a steady field of 20 microgauss in a few tens of billions of years.

However, there was a catch to everyone's models. It turned out that to match the observed field strengths, a galactic dynamo would have to start with a primordial field much larger than was plausible. For a while that halted progress. Meanwhile another voice was being heard.

When most of the crowd is cheering the favorite horse, there are usually a few contrarians that are betting that another horse will win. Russell Kulsrud is one of these independent minds. He pursued the primordial scenario tenaciously during the 1990s, while the dynamo seemed to gain more and more credibility. In recent years he's come around to offer a synthesis of the two theories.

Kulsrud's scientific credentials as a plasma physicist are impeccable, so his arguments command attention. He recently retired from a career at the Princeton Plasma Physics Laboratory, where he was helping to control thermonuclear fusion for electrical power generation. He also taught plasma physics at Princeton University for over twenty years. Astrophysics has always been something of a sideline for him.

Kulsrud thought that the primordial scenario had been discarded prematurely, with too little serious investigation. Beginning around 1986, he argued that fields we observe in the galaxy today could be fossils left over from its formation. He recognized the winding problem that Hoyle had raised, namely, that differential rotation in the disk would wind up a seed field far more than is observed. But he countered by pointing out that ambipolar diffusion would allow field lines to escape from the disk in a reasonably short time. In this way, he suggested, the buildup of magnetic flux in the disk and the winding of field lines would be limited.

In the decade of the 1990s, Kulsrud developed these ideas. He visualized the growth of the field in three stages. First, the protogalaxy collapses, increasing the intergalactic field strength by a factor of 10^4. Next, differential rotation winds up the field lines and the field strength increases. But the speed of ambipolar diffusion increases faster, in fact as the square of the field strength. Therefore, field lines diffuse more and more rapidly toward the top and bottom of the disk, where a portion of their lengths is ejected. So the field strength in the disk peaks and declines slowly. The field we observe today is only about a tenth of its earlier maximum.

Kulsrud and his student Armando Howard presented a heuristic model of this evolution in 1996. But their model explicitly ignored turbulence, which was unrealistic, and the theory received little support.

By 2007 Kulsrud had revised his ideas and presented them in a review of the subject, written with Ellen Zweibel of the University of Wisconsin at Madison. Now he visualizes the growth of a galactic field in a different set of three stages. First, the intergalactic field must arise from absolutely zero strength to a seed field of perhaps 10^{-13} to 10^{-9} gauss. Second, the seed fields are amplified to "dynamically interesting" strength, with which they can influence the motion of the gas. And finally, a dynamo can begin to operate to maintain the fields.

Several exotic processes are available to explain the first stage, and we will consider them in the next chapter. Much of the debate today concerns the second stage. The question is whether a dynamo begins to operate before or after a galaxy has formed. Most researchers favor the time after formation and during the lifetime of the galaxy. Kulsrud argues, in contrast, that an alpha-omega dynamo could not begin to function until the seed field was strong enough to influence the motion of the gases. Another mechanism, then, is required to amplify the primordial field.

The key to the problem, he maintains, is that you cannot create magnetic flux of one polarity from zero flux. (That would require the existence of isolated magnetic poles.) So positive and negative primordial flux must have been created in equal amounts. In order for a galaxy to end up with a predominant magnetic polarity, some flux of the opposite polarity must escape from the disk entirely. But frozen-in flux tubes can escape the disk to infinity only if they can shed their entrained mass of gas. And that requires that the field lines bow outward, to allow the gas in them to drain back into the disk. Therefore, the field strength must surpass a minimum strength. In short, a dynamo could start only after the primordial field has been amplified sufficiently by some independent mechanism.

Kulsrud favors some form of small-scale nonhelical turbulence to amplify the fields in the second stage. Several theories for this process have been advanced, including one by Kulsrud and colleagues in 1997, but just how the small scales would lead to galactic-sized fields is not entirely clear. It is fair to say that, at this point, the controversy hasn't subsided, but the dynamo paradigm still holds center stage.

Anvar Shukurov at Newcastle University, Newcastle upon Tyne, recently summarized reasons why this is so. He discussed three observational tests that mean-field dynamo theory supposedly passes and that the primordial scenario fails. The first is a prediction of the pitch angle of the field in spirals, which as we saw, ranges from 10 to 40 degrees.

In dynamo theory the pitch angle is determined by the competition between winding up the field lines and smoothing them by diffusion. With reasonable estimates for the rate of diffusion, a pitch angle of a few tens of degrees is easily obtained. Kulsrud's original theory, in contrast, would have the primordial field wind up rapidly to a maximum strength and then decay by diffusion over some 10 billion years to the present value. That theory would predict a pitch angle of about 1 degree. Kulsrud hasn't estimated a pitch angle with his latest theory, however, so it may be premature to say his theory fails on this count.

In a second test, dynamo theory predicts that the direction of the field is the same above and below the galactic disk. Shukurov argues that the primordial scenario would predict opposite directions, if the field survived at all. This is a weak test, though, because observers do not agree on the directions of the field on opposite sides of the disk.

Finally, Shukurov points to successful predictions by dynamo models of field *reversals* in the Milky Way and in M51. This may indeed be difficult to accomplish in a primordial field scenario.

If Rainer Beck and his international consortium are able to build their cherished Square Kilometer Array, they would be able to test dynamo theories in different galaxies. Meanwhile, theorists are tackling problems in several areas. The role of radial flows in maintaining the spiral fields is just beginning to be explored, for example. Magnetic fields in irregular and barred galaxies, winds from galactic disks, and the interplay of cosmic rays and magnetic fields are all hot topics. Stand by for some interesting science!

SOMETHING FROM NOTHING

Seed Fields

WE COME AT LAST to the vexing question of how the first magnetic fields in the universe arose. We've seen how magnetic fields can be *amplified* in many different astrophysical situations. Gravitational collapse is a prime mechanism. As a protostar forms from the interstellar gas; as a galaxy forms from the intergalactic medium; as a white dwarf or a neutron star shrinks to its present size—in all these events the magnetic field in the original medium is concentrated and therefore amplified. Then in Sun-like stars and in mature galaxies, a hydromagnetic dynamo can maintain or regenerate the field.

But how were magnetic fields created in the early universe? If we looked back far enough in the history of the universe, could we discover when magnetic fields first appeared?

These are some of the questions that astronomers have in mind as they probe for magnetic fields in the gas *between* galaxies, the so-called intergalactic medium. This is the new frontier. Observers are using a variety of techniques in their quest to discover and measure the weakest of fields in the most tenuous of gases. Only after they have a firm grasp of the size and structure of these fields can theorists attack the central question of origins.

Many theoretical proposals have been offered already, based on the fragmentary data. In this chapter we'll survey a few of them. First, we'll need some background on the large-scale structure of the universe. Then we can ask what the observers have to tell us.

LOOKING BACKWARD

The deeper we look into space the further back in time we see. The reason, as every schoolchild must know by now, is that light travels at a finite speed, 300,000 km/s. Consequently, the more distant a source is, the longer it takes a light beam to reach us. When light arrives from a distant galaxy, we see it as it was billions of years ago, when its photons started out on their incredible journey.

Therefore, we measure distances in light-years. The farthest quasar discovered so far lies about 13 billion light-years from us. It must have been born less than a billion years after the Big Bang.

Space has been expanding ever since, and as Edwin Hubble discovered, the farther away a galaxy lies, the faster it recedes. Astronomers are using this relationship now to construct a three-dimensional model of the distribution of galaxies. They've made a remarkable discovery: that even on the largest scales, the universe is not uniform, as Einstein postulated—it has a discernable structure.

As early as the 1770s, observers like William Herschel and Charles Messier recognized that the sky contains distinct patterns. They noticed that what they called nebulae and what we now know as galaxies, *cluster* on the sky. One of the richest clusters, visible in a small telescope, lies in the constellation Virgo. In the nineteenth century, when astronomers were able to measure the velocities of individual galaxies in this cluster, they concluded that the clustering on the sky is not accidental. Rather, the galaxies in the cluster are bound together by their mutual gravity. We are looking at a gigantic swarm of galaxies, perhaps 2,000 of them, which were presumably formed from the same enormous cloud of gas, eons ago. From the speed of recession of the center of the cluster, astronomers have determined its distance, about 60 million light-years from Earth.

Large clusters have also been found in the constellations Fornax, Centaurus, Leo, Perseus, and Coma Berenices. George Abell, a student at Caltech, compiled a catalog of 2,700 clusters in the late 1950s by scanning plates from the Palomar Sky Survey. The latest catalogs list about 5,000 clusters.

Our Milky Way Galaxy belongs to a local group (or subcluster) of six galaxies that includes the Andromeda Nebula, at a distance of 2.4 million light-years. The local group, in turn, is a member of the so-called Local Super-cluster, which includes the Virgo Cluster and groups in Leo, Ursa Major, and Draco. This supercluster, a loose collection of gravitationally bound objects, spans a distance of about 140 million light-years. The nearest superclusters to ours lie in the constellations Coma Berenices and Perseus, at distances of about 300 million light-years.

This was about all we knew of the structure of the universe as of 1975. In that year, astronomers at the Center for Astrophysics (CfA) at the Harvard Smithsonian Institute undertook an enormous task. They set out to deter-mine the positions, in three-dimensional space, of tens of thousands of gal-axies in our neighborhood. They had the technical advantage of digital im-aging, which allowed them to determine a precise angular position on the sky of hundreds of galaxies in each image. That provided two of the three coor-dinates in space. For the third coordinate, a galaxy's distance from Earth, they relied on a coarse spectrum of the galaxy. From its spectrum they could determine the galaxy's speed of recession, which is a measure of its distance. The whole process could be automated. These astronomers were data-mining the universe!

In 1989, Margaret Geller and John Huchra, two members of the CfA team, announced some of the first results of the survey. Their discovery stunned the community of extragalactic astronomers. They had uncovered the Great Wall, a thin sheet of clusters that extends for 500 million light-years across the sky, at a distance of about 200 to 300 million light-years. It was the largest coherent structure ever seen in the universe. Moreover, they showed that the sheet was the boundary of an immense void in which few galaxies could be detected. Evidently, the universe has a foam-like structure, in which galaxies collect on the surfaces of gigantic empty bubbles. This discovery completely upset the old idea that galaxies are uniformly distributed in space.

Other groups have since undertaken similar deep-sky surveys. In 1997 the Two Degree Field Galaxy Redshift Survey was begun at the Anglo-Australian Telescope in Australia. And in 2000 the Sloan Digital Survey was launched. Three years later, John Gott III and Mario Juric at Princeton University dis-covered the Sloan Great Wall of galaxies in the survey's data bank. It spans 1.4 billion light-years at a distance of 1 billion light-years and is now consid-ered the largest single object known.

Many scientists are intrigued by the foam-like structure of the universe and are busy trying to explain its origin. Among them are observers who are exploring magnetic fields at the largest scales. They have many questions in mind. Does a primordial field exist in the intergalactic medium, far from any superclusters? If so, does it provide the seed field that galaxies capture as they form? Or do the galaxies generate a seed field themselves as they coalesce? Could magnetic fields influence the way galaxies form?

These are questions that future observations of the early universe may be able to answer. For the present, observers of cosmic fields are focused on studying the fields in relatively near clusters and superclusters. These fields may be dynamically important, as they could provide additional pressure support to the gas between galaxies, as well as confining cosmic rays.

The Coma Supercluster and its richest member, the Coma Cluster, are favorite targets. As we shall see, they've yielded some important clues on the nature of cluster magnetic fields.

CLUSTERS, LIKE ANGELS, HAVE HALOS

Optical astronomers are well acquainted with the Coma Cluster. From its recession velocity, they've estimated that it lies at a distance of about 350 million light-years. It contains around a thousand galaxies that swarm in a sphere about 3 million light-years in diameter. The galaxies are mostly elliptical, with rather few spirals.

In 1959, a powerful radio source was discovered in the cluster. It was huge. If we could see it with our naked eyes, it would easily cover the full Moon. Nobody followed up for a decade, but in 1970 M. A. G. Willson used the Cambridge University's One-Mile Telescope, an interferometer, to scan the cluster at high resolution. He determined that at least eleven individual galaxies in the cluster were radio sources. But in addition, the whole cluster was embedded in a huge diffuse cloud, which he called a radio halo. It was the first of its kind and opened a path toward the exploration of magnetic fields inside clusters.

Willson identified the emission as synchrotron radiation from its characteristic spectrum. Synchrotron radiation, you will remember, is the energy that electrons emit as they spiral around the lines of force of a magnetic field.

To estimate the strength of the halo's field, Willson adopted the equipartition principle that had been introduced by Geoffrey Burbidge, a Cambridge astrophysicist, some years before. It amounts to assuming that the total energies of the radiating electrons and the magnetic field are equal. With this strong assumption, Willson derived a field strength of 2 microgauss for the halo.

Now this was a surprisingly large field. After all, the field strength in the disk of the Milky Way was known to be a few microgauss. One could reasonably expect that, in a tenuous halo of a cluster of galaxies, the field would be much smaller. Perhaps the equipartition assumption was at fault. Or possibly the Coma Cluster is special in some way.

In the following year, astronomers got a new look at the Coma Cluster. The Uhuru satellite discovered that the cluster possesses a soft x-ray halo that coincides with the synchrotron halo. From the x-ray spectrum, Herbert Gursky and his team at the U.S. Naval Research Laboratory determined that the halo's emission was thermal, at a plasma temperature of 70 million kelvin. So the halo emits both thermal and nonthermal (synchrotron) radiation.

All through the 1970s and 1980s, different observers measured the synchrotron radiation of the Coma Cluster. Their estimates of the halo's magnetic field varied widely, because the angular resolution of their observations varied. In 1978, for example, E. Valentijn, a Dutch astronomer, calculated the equipartition field strength as 0.4 microgauss, five times smaller than Willson's number.

In 1989 a team of Canadian astronomers led by Phillip Kronberg at the University of Toronto measured the field strength in the Coma Cluster with more precision than anyone before them. They employed a familiar tool, Faraday rotation. In chapter 11 we saw how radio astronomers use observations of Faraday rotation to estimate the magnetic fields in the Milky Way.

Kronberg and company observed the Coma Cluster at high resolution with the Very Large Array (VLA) and the Dominion Radio Astrophysical Observatory's telescope. They measured the Faraday rotations of eighteen distant sources behind the Coma Cluster and combined them with x-ray data to determine the average halo field strength. It worked out to 1.7 microgauss, with an uncertainty of 0.9 microgauss. So the team reached an important conclusion: Coma's halo fields really are as strong as equipartition estimates suggest.

But the field strengths one measures seem to depend on the data one starts out with. In 1999, Roberto Fusco-Femiano and his large Italian team discovered *non-thermal* x-rays in the Coma Cluster halo, with photons as energetic as 25 keV. They combined their data with radio observations of the cluster and derived the average field strength as only 0.15 microgauss. It may be possible to reconcile the different estimates if one supposes that the field varies considerably throughout the halo, not an unreasonable possibility.

STRONG FIELDS EVERYWHERE

With all the attention paid to the Coma Cluster, one might ask whether it is typical of clusters. Phillip Kronberg and his student Tracy Clarke set out in 2001 to answer that question with new observations at the VLA. They selected sixteen clusters that varied widely in x-ray brightness and had normal or weak radio halos. Then they observed one or more background sources through each halo and determined their rotation measures. Combining their radio data with x-ray observations from the ROSAT satellite, they calculated halo magnetic field strengths in the range of 5 to 10 microgauss. These halo fields were even stronger than those of the Coma Cluster!

From their data they could also calculate the total magnetic energy of a typical halo. With a field of 5 microgauss, and a halo diameter of 1.5 million light-years, the total energy works out to 10^{48} kilowatt-hours, or about 2 percent of the halo's thermal energy. Now that is a staggering amount of energy and cries out for an explanation.

Kronberg and Clarke announced their results at a meeting of the American Physical Society in April 2000. It caused a sensation. Russell Kulsrud, the proponent of primordial fields whom we met in chapter 11, was skeptical that the halo fields could be so strong but agreed that, if confirmed, they could be very difficult to explain. Eugene Parker, the pioneering theorist of magnetic fields, was intrigued by the huge total magnetic energies. And Stirling Colgate, a theoretical physicist at Los Alamos National Laboratory, immediately suggested a source for the energy.

Colgate was an imaginative theorist who had advanced some controversial ideas concerning the dynamics of supernovas. He had predicted, for instance, that a blast of neutrinos would help to drive the shock wave from a supernova. His predictions were confirmed in 1987, when neutrinos were detected from a massive supernova in the Large Magellanic Cloud. We'll return to his ideas about cluster fields in a moment.

WHAT LIES BETWEEN CLUSTERS?

If the field strengths in cluster halos are as large as a few microgauss, perhaps they don't represent the primordial fields *between* clusters. In 1990 Kronberg joined a team of radio astronomers from the University of Bologna to find out. The team, led by Gabriele Giovannini, made a detailed map of the Coma Supercluster with the Westerbork Synthesis Radio Telescope in Holland. This supercluster consists of the Coma Cluster; the cluster Abell 1367; and a third cluster with the clumsy label 1253 + 275. The team discovered a faint luminous "bridge," at least 130 million light-years in length, which connects the Coma and 1253 + 275 clusters.

Now the diameters of the individual clusters are "only" about 3 million light-years. Therefore, the team argued, this bridge represents a portion of the intercluster medium from which the clusters formed. Assuming the emission is synchrotron radiation, they calculated the equipartition field strength as 0.3 microgauss. Once again, a remarkably strong field had been found.

In 2006 Kronberg returned to this question of intercluster fields. With his student Yongzhong Xu and two other colleagues, he selected two superclusters. The Hercules Supercluster lies at a distance of about 450 million light-years and contains twelve clusters. The Perseus-Pisces Supercluster contains about fifteen clusters at about 200 million light-years.

They collected Faraday rotation measures of sources behind each supercluster and in control regions nearby. They found that rotation measures in the superclusters were larger than in the control regions. But the Milky Way, which lies in the foreground, introduced uncertain contributions to the measures and prevented the team from obtaining unambiguous results. The best they could do was set an upper limit of 0.3 microgauss to the intercluster fields.

Kronberg doesn't discourage easily, though. He's been on the trail of intergalactic magnetic fields since the mid-1970s and is determined to find some final answers. So in 2007, he and his team set up a joint observing program, using the 300-meter Arecibo radio telescope and the radio interferometer at the Dominion Radio Astrophysical Observatory in Canada. This time they returned to the Coma Supercluster, which is nearly free of fore-ground contamination by the Milky Way.

They mapped the synchrotron radiation in a region about 45 million light-years in diameter that was centered on the Coma Cluster. They found a gold mine: patches of diffuse emission fainter than any seen before. In one such diffuse region, 12 million light-years long, they detected magnetic field strengths of 0.3 to 0.4 microgauss. Once again microgauss fields seemed to be the rule.

In addition there were long streaks of "cirrus" that contain no galaxies. The team suggested this glow was produced by cosmic rays trapped in a weak magnetic field. But exploring those fields would require even better observations.

FIELDS IN THE EARLY UNIVERSE

The Coma Cluster, for all its productivity, lies only 300 million light-years away, at a redshift of $z = 0.0231$ (note 12.1). How far away, or equivalently, how early in the life of the universe, do we have to look to see the first magnetic fields?

To answer this question, Kronberg and two colleagues observed the Faraday rotation of a distant quasar with the VLA in 1992. This quasar has a redshift of $z = 1.081$, which corresponds to a distance of 7 billion light-years. They discovered that the Faraday rotation did not arise near the quasar but in a foreground galaxy at a redshift of $z = 0.395$. Using auxiliary spectra of the galaxy, they determined it was a normal bisymmetric spiral. The rotation measures indicated its field strength lies between 1 and 4 microgauss, not too different from that of the Milky Way.

Kronberg assumed that a dynamo in the galaxy had amplified a tiny primordial field to the microgauss level. He was able to estimate the strength of the seed field that the dynamo required to reach this level during the finite life of the galaxy. With different assumptions on the time of formation, he found initial seed fields ranging from 10^{-23} to 10^{-18} gauss. (In scientific notation, 10^{-18} is equal to the decimal 0.000,000,000,000,000,001.) Not exactly tight limits, but at least an attempt to test previous theoretical estimates.

One would prefer to look for magnetic fields in more distant quasars. The farthest nine quasars detected so far in the Sloan Digital Survey have red-shifts larger than z = 6, but none of these emits synchrotron radiation, which is necessary for the detection.

In 2006, however, a group of scientists from Columbia University identified a radio-emitting quasar with a redshift of z = 6.12. Exactly when this object emitted the radio waves depends on the cosmological model you choose, but a reasonable estimate is 900 million years after the Big Bang, or roughly 12.6 billion years ago. Synchrotron radiation requires magnetic fields; therefore, they existed at least this early. So far we have no measure of their strength.

So to summarize, observers have found field strengths of about a microgauss in the diffuse gas within a cluster as well as between clusters. Such field strengths are similar to those in fully formed galaxies and correspond to a small percentage of the thermal energy of the gas. The intercluster fields are also well ordered over distances of least 300,000 light-years. Finally, the seed fields from which these fields were derived could be weaker than 10^{-18} gauss.

These are some of the constraints that theorists face in trying to explain the origin of magnetic fields in the universe. We'll see that many ideas have been proposed. A few have been rejected; none have been confirmed so far.

ON THE ORIGIN OF MAGNETIC FIELDS

Taking the first step, from zero fields to a small seed field, is the most difficult. As we have seen earlier, dynamos in stars or galaxies offer a plausible way to *amplify* a seed field, but one must start with a field, however weak. Many researchers invoke a mechanism invented, or discovered, by Ludwig Biermann in 1950. We've met Biermann several times in past chapters. This shy German astrophysicist contributed several seminal concepts, including the existence of the solar wind. The so-called Biermann Battery was one of his most influential ideas. To his credit, it is still relevant after fifty years. Here's how it works.

Imagine a rectangular sheet of plasma, with the same shape as a page in this book. Suppose the plasma temperature decreases from right to left along the width, with a steeper gradient near the bottom of the sheet than near the top. Such a distribution of temperature could occur in a shock within the primordial gas, for example.

Now consider the forces on a particular electron and a nearby proton at the top edge of the sheet (fig. 12.1 top). Each particle feels the force of gravity pulling it to the right, where most of the mass lies, but the gravitational force on the electron is much smaller than on the proton, because its mass is so much smaller.

Both particles feel the same amount of force to the left from the gradient of the temperature, however (note 12.2). As a result, the electron would feel a larger *net* force urging it toward the left than the proton and would accelerate away from the proton, except for their mutual electrostatic attraction. The electric field (E) between the particles keeps them loosely coupled.

Near the bottom of the sheet (fig. 12.1 bottom), the gravitational forces on electrons and protons are the same as they are at the top edge. And both particles feel the same amount of force from the gradient of temperature. But the gradient is larger here, so the electric field (E) required to keep the particles coupled is also larger.

We are home free: Faraday's law of induction tells us that parallel and unequal electric fields will drive an electric current in a closed circuit between the top and bottom of the sheet. And the current will generate a magnetic field. As long as the electric fields remain unequal, the magnetic field

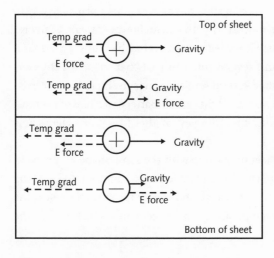

Fig. 12.1. A sketch to illustrate the Biermann Battery. In both upper and lower panels, the plasma temperature decreases from right to left, with a steeper gradient in the lower panel. The force due to the temperature gradient is shown as a dotted arrow, labeled "Temp grad." Gravity pulls both proton and electron to the right in this example and is shown with solid arrows. The unbalanced electric force between top and bottom panels will drive an electric current and, hence, create a magnetic field.

will *continue* to grow. In the setup shown in the figure, the field would point out of the paper.

Russell Kulsrud constructed a model for the origin of a primordial field that depends critically on the Biermann Battery. He and his colleagues proposed in 1997 that magnetic fields are generated in three stages in regions about to collapse into galaxies. In the first stage, which they simulated with a hydrodynamic computer code, density variations in the cosmic hydrogen gas induce flows, which develop into shocks. The shocks are like those thin plasma sheets we described just now and are ideal sites for the Biermann Battery to function. Kulsrud and friends estimated that by an age of the universe corresponding to $z = 3$ the field would build to 10^{-21} gauss. (That is, one-quadrillionth of a microgauss!)

In the second stage, vortical motions develop in the shocks and the gas becomes highly turbulent. The smallest eddies amplify the field by winding it, and they continue until the energy of the field equals the turbulent energy of the eddy. At this point, Kulsrud estimated, the field could be as large as 10 microgauss, but coherent only over the size of the smallest eddies. In the third stage, the field would presumably spread uniformly over the whole galaxy and decrease in strength to the submicrogauss level, but Kulsrud's team was unable to follow this process numerically.

George Davies and Lawrence Widrow, two physicists at the University of Toronto, developed this model further. They drew attention in 2000 to the spherical rarefaction wave that would propagate outward as the cosmic gases collapsed into a protogalaxy. The wave accelerates the neutral gas ahead of it so that it can fall into the protogalaxy. As the gas crosses the wave, it changes direction and generates vortical motions, which the scientists could follow in detail. In addition, they employed the Biermann mechanism to create a magnetic field in the shock. Turbulence plays no role in their model. They estimated that the vortical motions would create a seed field of about 10^{-17} gauss, 10,000 times that of Kulsrud and company.

The first generation of stars is another attractive location for the creation of seed fields.

As a protostar contracts, the same combination of shocks and vortical motions appear as in protogalaxies, only on a smaller scale. Once again the Biermann Battery is the essential mechanism. The very weak fields that appear could be amplified in the type of accretion disk dynamo we described in chapter 8. Then either stellar winds or jets that occur during star formation could disperse the fields into intergalactic space.

As we discussed earlier, however, a protostar cannot collapse without shedding its angular momentum. That requires an accretion disk with an embedded magnetic field. So we have a chicken-and-egg situation here: no star formation without a preexisting field and no generation of a field without a star in formation. Nobody, as far as I know, has solved this dilemma so far.

The first generation of stars in galaxies could generate magnetic fields *indirectly*, however, by re-ionizing the surrounding primordial medium with their ultraviolet radiation. A spherical ionization "front" would propagate from each star, converting the neutral gas ahead of it to plasma. The front could be another possible site for the Biermann Battery to function.

Nick Gnedin and colleagues at the University of Colorado simulated this process in considerable detail. They followed the development of several fronts as they grow and overlap. In the process, magnetic fields grow from zero to about 10^{-19} gauss, a respectable seed field. Moreover, the intrinsic size of the fronts translates into a field that is well ordered over distances of several million light-years, as is observed in the intercluster medium.

Yet another possible site for the creation of magnetic fields is the accretion disk around a supermassive black hole at the center of a galaxy. Stirling Colgate was convinced that only such a process could account for the enormous energies Phillip Kronberg had found in cluster halos.

In 2000 Colgate and two colleagues published a paper with the grandiose title "The Origin of the Magnetic Fields in the Universe: The Plasma Astrophysics of the Free Energy of the Universe." They proposed that intergalactic fields are not primordial but are ejected from black hole accretion disks. They worked out the details of an alpha-omega dynamo that could operate in the accretion disk and showed how a seed field could be amplified far beyond the microgauss levels that are observed in cluster halos.

Kronberg was intrigued. He was eager to collaborate with Colgate, and in 2001 they published an analysis of two types of radio sources. At one extreme were the giant double-lobed sources like Cygnus A, which are, without doubt, powered by the supermassive black holes at their centers. At the other extreme were smaller sources that are located at the centers of dense clusters.

Estimates of the minimum energy content of the double lobes (of order 10^{48} kilowatt-hours) suggested that the black holes must convert gravitational energy into magnetic energy with at least 10 percent efficiency. The expanding lobes, they wrote, would be major sources of the intergalactic field. Similarly the galactic black holes in clusters would generate fields that fill the cluster halo.

Other theorists, like Ralph Pudritz, have modeled the amplification of magnetic fields in an accretion disk around a massive black hole. Abraham Loeb and others have described how the fields are then ejected from the young galaxy by winds or jets into the intergalactic medium. But nobody that I have found has described how the seed fields are generated *near* the hole, by the Biermann mechanism or other means.

We turn now to scenarios of the creation of magnetic fields in the early universe, before galaxies formed. The physics involved is generally well beyond the scope of our discussion, so we'll just describe a few of the critical transitions in the life of the universe, the conditions that prevailed, and the general ideas behind some of the scenarios.

According to the current "standard model" of cosmology, the universe went through an interval of explosive inflation after the Big Bang, in which the radius increased by a factor of 10^{50} or more. After inflation ended at an age of 10^{-32} seconds, the universe was filled with an enormously hot soup of quarks, "gluons," and photons. With further expansion, the universe cooled to about 10^{13} K. Between perhaps 10^{-6} seconds and 1 second, electrons and neutrinos formed, and quarks joined to form protons and neutrons. This event is the so-called quantum chromodyamics (QCD) transition. Young electrons roamed freely through the gas and scattered photons so effectively that the photons were tightly coupled to the electrons.

Then between one second and three minutes, at a temperature of about 10^{10} K, protons and neutrons combined into nuclei of deuterium, tritium, and helium. This "matter-dominated" universe continued to cool. At an age of 380,000 years and a temperature of about 3,000 K, electrons and nuclei combined into atoms of hydrogen and helium, in a process called recombination. Photons were able then to decouple from matter and move independently. They became the cosmic microwave background we observe now. At this stage, the universe was relatively cold and dark.

Around an age of 500,000 years, fluctuations of density in the gas triggered gravitational collapse, and the first protogalaxies were born. Quasars were among the first to form. Soon thereafter the first stars appeared. Their strong ultraviolet light re-ionized the gas around them. Finally, dwarf galaxies coalesced to form larger galaxies like the Milky Way.

So to summarize, there were four important transitions that concern us: the end of inflation, when electrons and neutrinos appeared; the QCD transition, when protons and neutrons formed; recombination, when light atoms were created; and re-ionization, when the gas near stars was converted to plasma.

Could magnetic fields have originated immediately after the Big Bang, during inflation? A few brave souls have explored this possibility. Inflation attracts theorists because it is able to distribute an initial field throughout the universe, which would solve one pesky problem. But that advantage also has a downside: inflation would dilute any initial field so severely as to push the field strength toward zero. Actually this isn't a fatal difficulty, because even a seed field of 10^{-21} gauss would be acceptable.

To avoid this difficulty entirely, theorists have invoked a variety of exotic particles and fields (such as the "dilaton") to *amplify* the field during inflation. The final predicted field strength therefore depends a lot on the assumptions they build into the theory. Without some way to test such theories experimentally, they remain interesting exercises.

The physics of the universe is much better understood following the end of inflation, so several theorists propose to create magnetic fields just after the QCD transition. Charles Hogan, a Caltech physicist, was the first in 1983. He pictured protons and neutrons forming from quarks in gigantic bubbles throughout the primordial soup. As the bubbles expand and collide, shocks are formed at the walls of the bubbles. He assumed that the Biermann mechanism would create a tangled seed field in the shocks and that a dynamo would appear that would amplify them. He also imagined that reconnection of the field lines would occur when two shocks collided, so that the field could become untangled. His ideas were innovative but not fully developed.

Günter Sigl and colleagues at the University of Chicago took a much closer look at the mechanisms at the walls of the bubbles. They showed in 1997 that the gradient of temperature across a shock causes different rates of diffusion by electrons and protons. A separation of electric charges occurs, currents flow, and these produce seed magnetic fields. This mechanism is similar to the Biermann Battery.

Sigl and company estimated that just after the QCD transition, the field strength could be as large as 10^7 gauss, in bubbles a meter in diameter. As the universe continued to expand, the field strength would decrease. By the time of recombination (at an age of 380,000 years), the field strength would have dropped to 10^{-25} gauss in bubbles 300,000 light-years in diameter. Protogalaxies could form soon afterward, immersed in this weak seed field. So this model also solves the difficult problem of spreading coherent magnetic fields over distances larger than a galaxy.

The 380,000-year interval between the QCD transition and recombination also holds promise for generating the first magnetic fields. A group of Japanese scientists, led by Kiyotomo Ichiki at the National Astronomical Observatory of Japan looked at this possibility in 2006. They pointed out that photons would scatter off electrons much more frequently than off protons. Therefore, electrons and protons would move with different speeds, which would generate currents. The currents, in turn, would produce magnetic fields. The researchers estimated a fields strength of 10^{-18} gauss, coherent over 3 million light-years, at the time of recombination. According to their theory, the generation of primordial fields would end then, because electrons would be bound into atoms. And the seed field strength would decrease to 10^{-24} gauss at the present time.

This predicted field is far too weak to be detectable with polarization measurements. So Ichiki and friends suggested another possible method. When a massive star explodes, it accelerates electrons to relativistic energy and the electrons radiate gamma rays. Ichiki suggests that the gamma ray photons would be delayed relative to photons of visible light, because the seed fields would deflect the radiating electrons. Perhaps so, but such a measurement must await a future space telescope.

We've seen that astrophysicists have concocted a variety of schemes to account for the primordial magnetic fields. The fields could be generated at different stages in the life of the universe and in different locales. There doesn't seem to be a problem *generating* a tiny seed field, although estimates differ on its possible size. Scenarios also exist for *amplifying* a seed to the microgauss strengths we observe, say, in halos. The real problem is to account for the *coherence,* or uniformity, of these large-scale fields, because they can become disordered by stellar births in as little as 30 million years. Time will tell.

I hope I have been able to introduce you to some of the arenas in which magnetic fields play a role. In closing, I'd like to remind you of your own magnetic nature. Every molecule of water in your body contains two hydrogen atoms whose central protons possess "spin." Spin is a fundamental property, like charge or mass, and is associated with a minute magnetic field. So you are permeated with tiny, disordered magnetic fields. In fact medical diagnosticians use this property to image parts of your body with nuclear magnetic resonance imaging. But I hope you will never find this necessary.

NOTES

1. GETTING REACQUAINTED WITH MAGNETISM

1.1. Physicists now recognize four basic forces in nature: gravity; electromagnetism; the "strong" force binding the nuclei of atoms; and the "weak" force that governs radioactivity. During the first microseconds after the universe was born in a Big Bang, it is thought, all four forces were joined in a superforce. As the universe cooled and expanded to its present state, the forces separated.

1.2. Ampère discovered that two long parallel wires that carry current in the same direction *repel* each other. Therefore, one way to define the ampere, the unit of current, is to relate it to a measurable mechanical force, the kind one exerts against a closed door, for example. If the two wires each carry one ampere of current, and are separated by 1 meter, the force of repulsion will amount to one newton, or 0.22 pounds.

Then we can define the unit of magnetic force, or more precisely the flux density, as that which appears at a distance of 1 meter from a long wire carrying one ampere of current.

The unit is called the tesla, after Nikola Tesla, the Serbian-American inventor. At the magnetic poles of the Earth the flux density is about 20,000 times smaller than a tesla, or about 0.5 gauss, in older units. One tesla equals 10,000 gauss. In general, astronomers prefer to use the gauss.

2. THE EARTH

2.1. In 1842, Ross commanded two ships in an exploration of the east coast of Antarctica, where he discovered the ice shelf and sea named after him. Sir Douglas Mawson located the South Magnetic Pole in January 1909. He described his ordeal in *The Home of the Blizzard* in typical matter-of-fact style.

2.2. The Earth's heat has two sources: the primordial heat acquired during its formation and the continuous heating due to radioactive decay. Hot cells would rise buoyantly, release their heat, and sink to pick up another load.

2.3. The Coriolis force is a fictitious force that appears because we are located on a rotating sphere. A bullet fired due north from the equator seems to an observer in midlatitudes to drift to the east. She would claim a force deflected the bullet, but the drift is actually caused by the faster eastward speed of the gun at the equator, relative to midlatitudes. Similarly, the omega loop on the Sun experiences an apparent force as it rises.

2.4. Werner Heisenberg, one of the founders of quantum theory, was asked on his deathbed what he would ask God when he saw him. "I would ask Him why there is turbulence," he replied.

3. SUNSPOTS AND THE SOLAR CYCLE

3.1. This quote was taken from Helen Wright's fine biography of Hale, *Explorer of the Universe*.

3.2. The spectrum lines of an atom split into a pattern of three or more when the atom is embedded in a magnetic field. In the simplest case, a triplet, the central component is linearly polarized transverse to the field direction and the two outer components are circularly polarized, one clockwise, the other anticlockwise.

3.3. Granules are typically 1,500 km in diameter, bright in their centers, and surrounded by a dark lane only a few hundred kilometers in width. Plasma in a granule rises, spreads horizontally with a speed of about 1 km/s and sinks in the lane with speeds of several kilometers per second. Vortices and enhanced magnetic fields have been detected in the lanes. The granule fades, or breaks up, within about ten minutes.

Supergranules are also a type of convection cell. They are 30,000 km in diameter on average and persist for about a day before dissipating. The plasma flows radially outward from the center at a fraction of a kilometer, sweeping any magnetic field toward the boundaries of the cell.

3.4. Plasma is a good electrical conductor. If a flow begins to move across magnetic field lines, a current is induced, which interacts with the existing field to generate a so-called Lorentz force. This force acts to restrain the plasma from any further movement. Therefore, the circular motions of a convection cell would be strongly inhibited in a strong magnetic field. If, however, the flow's kinetic energy density exceeds the field's magnetic energy density, the flow can distort the field, as in fig. 1.4.

3.5. Imagine a drunken man standing next to a lamppost. He takes a step in some direction, then an equal step in a randomly chosen direction, and so on. Over a long period of time, the average of all his locations will center on the lamppost, but his average distance from it will increase as the square root of the number of steps. He has taken a random walk.

3.6. The plasma in a rising magnetic tube will expand and cool as it reaches regions of lower pressure. If it arrives cooler than its surroundings, it will therefore be denser and will stop and then sink. If, on the other hand, it arrives hotter than its surroundings, it will be less dense and will be buoyant.

3.7. Gustav Spörer, a German astronomer, discovered that sunspots were extremely rare between 1420 and 1570. The existence of the Spörer Minimum was later confirmed in 1883 by British astronomer E. Walter Maunder, who found from old sunspot records that hardly any spots were visible from about 1645 to 1715, an interval of six normal solar cycles. This Maunder Minimum was accompanied by extremely cold weather in Europe, a fact that has stimulated much research among climatologists and solar physicists.

3.8. I should hasten to add that measurements of the strength of the field become increasingly difficult above the chromosphere. In the low corona, where we can capture images of x-ray loops, we can determine the direction but not the magnitude of the field. The coronal plasma density is too low and the temperature is too high for the usual spectroscopic methods to work. Nevertheless, observations with radio interferometers have given us some useful measurements. Alternatively, measurements in the chromosphere can be extrapolated numerically into the corona by methods we'll cover in chapter 4. Similarly, the strength, and particularly the shape, of the field below the surface can be determined only by indirect means and is still controversial.

3.9. Alan Title, a former student of Robert Leighton, is a talented experimentalist. He heads a research team at the Lockheed Palo Alto Laboratory, where he has designed and built a series of successful instruments for solar satellites such as SOHO and TRACE. I remember him showing me his latest imported car, a sleek Ferrari. Alan is shorter than I am and had no trouble sidling into the driver's seat, which sits about 6 inches from the pavement. I almost had to lie down to get seated. The ride was worth the effort, however.

4. THE VIOLENT SUN

4.1. Andrei B. Severny, solar physicist at the Crimean Astrophysical Observatory, was observing solar magnetic fields in the early 1960s. He was probably the first to point out the key role of the neutral line between antiparallel fields in the photosphere. Actually these fields are divided in three dimensions by a surface called the separatrix.

4.2. A charged particle moving across a magnetic field experiences a Lorentz force that is perpendicular to both the velocity and to the field direction. In general, the particle will be forced to spiral around the field lines. Only if the velocity is along the field direction is the Lorentz force equal to zero. Another way to express this result is to say that the cross-field resistivity is much higher than the resistivity along the field.

4.3. Hannes Alfvén and Per Carlqvist at the Royal Institute of Technology, in Stockholm, Sweden, proposed a theory of solar flares in 1967. They pointed out that a circuit in plasma could carry only a definite maximum amount of current before it "explodes." At that moment the magnetic field associated with the current collapses rapidly and induces violent heating. The plasma limit has been verified, but their concept of the field collapse was disproved.

4.4. Giovanelli was a tall, bald man with an infectious grin and a shy manner when I met him in the 1960s. He virtually sparkled with fresh ideas. For many years he supervised a talented solar group at the Commonwealth Scientific and Industrial Research Organization in Sydney, where a strong program in experimental and theoretical research was underway.

4.5. Parker was relying on the classical diffusion rates calculated in the 1950s by Lyman Spitzer, physicist at Princeton University. Laboratory experiments revealed later that diffusion rates could be much faster in real plasmas.

4.6. Many of these problems have been solved. The International Thermonuclear Experimental Reactor (ITER) is a large-scale fusion reactor that will be built in Cadarache, France, beginning in 2007.

5. THE HELIOSPHERE

5.1. The Alfvén speed of propagation V, is proportional to the magnetic field strength B and inversely proportional to the square root of the plasma density r. Or, $V = B/(4 \pi \rho)^{1/2}$.

5.2. The expansion of the solar wind has been compared to the flow of compressed gas through a conical De Laval nozzle, which is used in rocket engines. The nozzle shrinks down to a minimum diameter throat and then flares out. The gas is forced to pass through the throat, where its speed reaches the speed of sound. Beyond the throat the gas is supersonic. The critical height in the corona, where the wind speed reaches sound speed (or with a magnetized plasma, the Alfvén speed), would correspond to the throat of a nozzle.

5.3. In an Archimedes spiral the distance between successive turns is a constant. We'll encounter another type, the "logarithmic" spiral, in chapter 11.

5.4. During the 1970s, radio observations did detect high-speed winds at the solar poles. Barney Rickett and Charles Coles measured the turbulence in the polar corona by its effect on the scintillation of galactic radio sources as their radiation passed through the corona. Speeds as high as 700 km/s were deduced.

5.5. Waves can exert pressure. Sound waves in air press on our eardrums, for example. Even electromagnetic waves propagating in a vacuum exert pressure. They can be thought of as showers of energy packets (photons) that beat on a target. Alfvén waves in plasma are not essentially different.

5.6. R. E. Hartle and Peter A. Sturrock devised the first "collisionless two-fluid" model of the wind in 1968. They recognized that energy exchange between electrons and protons would be so slow in dilute plasma as to allow them to have different temperatures. They computed the independent heating rates of the two species and compared the resulting temperatures with observations near Earth. The predicted proton temperature was far too low. Extended heating of the wind was indicated.

6. THE EARTH'S MAGNETOSPHERE
AND SPACE WEATHER

6.1. Actually, Humboldt was not the first to see such magnetic fluctuations. In 1741, Anders Celsius, a Swedish astronomer, and Olaf Hiorter, his assistant, saw the effect during a bright aurora. Celsius also invented the temperature scale we use today, except that he assigned the boiling point of water as 0 degrees and the freezing point as 100 degrees. A Swedish manufacturer of thermometers inverted the scale, assigning 0 degrees for the freezing point.

6.2. Carrington discovered that the surface of the Sun rotates faster at its equator and slower near its poles. He also recognized that new sunspots emerge closer and closer to the Sun's equator in the course of its eleven-year cycle. Finally, he determined with high accuracy the inclination of the Sun's rotation axis with respect to the ecliptic.

6.3. Elias Loomis, a Yale University professor, published a map in 1860 of the frequency with which auroras are seen in northern regions. He showed that they are seen most often from an oval displaced from the magnetic pole. The IMAGE satellite has taken beautiful pictures of the spread of auroras around this oval during a magnetic storm.

6.4. James Van Allen was a cosmic ray physicist at the University of Iowa. During the International Geophysical Year (1957–1958), his team launched Explorer 1, the first American satellite built to explore outer space. The spacecraft carried a Geiger counter, a simple device that detects low-energy charged particles. (It is commonly used to detect radioactive materials.) Explorer 1 was able to transmit data during only a fraction of its flight. The data were puzzling, copious cosmic ray counts at low altitudes but zero counts at high altitudes, around 2,500 km. The puzzle was solved after Explorer 3 returned a tape recording of the data during the complete flight. The Geiger counter had saturated at high altitudes, indicating a very high flux of energetic particles. The radiation belts had been discovered.

6.5. In dense plasma, like the surface of the Sun, charged particles behave almost like a fluid with a well-defined pressure and temperature. In dilute, collisionless plasma like the magnetosphere, the charged particles can travel freely for long distances before encountering another particle. Therefore, they interact, not by colliding, but by emitting and absorbing a variety of electromagnetic waves that travel at the speed of light. In both collisional and collisionless reconnection, a large part of the available magnetic energy is converted into the kinetic energy of the charged particles. The process generates beams of nonthermal, high-energy electrons and protons as well as jets of hot plasma. The microphysics of reconnection is still a subject of intense research and debate.

6.6. In X-type reconnection, field lines of opposite polarity cross at a so-called neutral point. The resulting configuration resembles a capital X. The V-shaped reconnected lines separate at the neutral point. Each half now has opposite magnetic polarity.

7. THE PLANETS

7.1. Pluto was long accepted as the ninth planet, born about the same time as the others, but in 2006 was demoted to be a "dwarf planet," like other small rocky bodies in the Kuiper belt. Its origin, and that of Charon, its satellite, is debatable.

7.2. The Lunar Laser Ranging Experiment has racked up several invaluable scientific discoveries. Among them: Newton's universal gravitational constant, G, has changed less than 1 part in 100 billion in the past forty years; the Moon is drifting away from Earth at 3.8 cm/yr because of the Earth's ocean tides; and Einstein's assumptions for the general theory of relativity are valid.

7.3. In the thermoelectric effect a temperature difference across a metal bar causes a voltage difference to arise. The converse is also possible; voltage differences create temperature differences. Electrochemical effects convert chemical into electrical energy, reversibly, as in a charging or discharging battery.

7.4. Schiaparelli realized that his reports were being seriously overinterpreted. When he saw Lowell's drawings of canals on Mars, he concluded that most were imaginary. Later research suggests the canals are visual illusions, caused by the brain's tendency to "connect the dots."

7.5. The ESA launched the Mars Express spacecraft in June 2003, and it is performing well as of 2007. For example, its radar discovered a vast deposit of water ice at Mars's south pole in March 2007. It carries a variety of instruments to map the composition of the surface and atmosphere, but unfortunately no magnetometers.

7.6. Seafloor spreading was one of the important clues that led to the acceptance of the theory of continental drift. According to the theory, Earth's crust is composed of giant tectonic plates that are continually growing as lava spews from the mantle at midoceanic ridges. In 1963 Frederick Vine and Drummond Matthews discovered magnetic bands of alternating polarity parallel to a ridge. They explained that the Earth's magnetic field reverses polarity periodically and that these changes are frozen into the spreading lava, creating the bands. The same mechanism may not apply to Mars, however.

8. MAGNETIC FIELDS AND THE BIRTH OF STARS

8.1. Leon Mestel quoted this comment in *ASP Conference Series* 248, 3, 2001.

8.2. Amino acids and formaldehyde are some of the more exotic molecules detected in the interstellar medium.

8.3. For those of you who are mathematically minded, the critical mass M_m for a cloud with field strength of B microgauss and a density of n hydrogen atoms/cm^3 is $M_m = 4.4 \times 10^4 \, B^3/n^2$, in solar masses.

8.4. The neutral molecules collide with the protons and tend to follow their motion. This is a type of frictional force on the neutrals.

8.5. BIMA has recently merged with Caltech's Owens Valley Radio Observatory to form a new observatory, the Combined Array for Research in Millimeter-wave Astronomy (CARMA). The array will be relocated to a high-altitude site in eastern California.

8.6. Alfred Joy discovered T Tauri in 1945 at the Mount Wilson Observatory. He identified it as a new class of variable star because of its wild fluctuations in brightness and its peculiar spectrum. We now know that T Tauri flashes at all wavelengths, from x-ray to infrared.

9. ABNORMAL STARS

9.1. The Harvard spectral sequence of normal stars is essentially a scale of surface temperature: O (30,000–60,000 K) Blue; B (10,000–30,000 K) Bluish; A (7,500–10,000 K) Blue-White; F (6,000–7,500 K) White; G (5,000–6,000 K) Yellow; K (3,500–5,000 K) Orange; M (2,000–3,500 K) Red.

The spectrum of a star varies with its surface temperature. Thus, for example, the hot O stars have spectral lines of helium, while G stars have none but do have strong lines of sodium and calcium.

9.2. If a star rotates with its axis exactly along the line of sight, all parts of its visible surface move transverse to the axis. Therefore, the radiation from each part is not Doppler shifted and the contributions of all the parts add up at the exact center of the spectral line, which is therefore "sharp." If the axis is tilted with respect to the line of sight, however, each part has a component along that line and its radiation is Doppler shifted. The sum of the radiation from all the parts is, then, a broadened line.

9.3. In 1852 George Gabriel Stokes, an Irish mathematical physicist, introduced a way of decomposing the polarization state of an electromagnetic wave, such as light. Four parameters are required, labeled I, Q, U, and V. I measures the total intensity of the light; Q and U measure the strength of the linearly polarized components in perpendicular directions; and V measures the strength of the circularly polarized component. Stokes also contributed many important results in hydrodynamics. He graced the Lucasian Chair in Mathematics, once held by Isaac Newton at Cambridge University, for many years.

9.4. The interior of the Sun is filled with a complicated pattern of standing sound waves with periods around five minutes. The waves are visible at the surface as a pattern of patches that rise and fall periodically. Solar astronomers have analyzed these surface patterns and derived from them the temperature profile and rotation behavior of the solar interior. This subject is called helioseismology, in analogy with the study of terrestrial earthquakes. See chapter 3.

10. COMPACT OBJECTS

10.1. An international team of astronomers combined new observations of Sirius B in 1998 and calculated that its surface temperature is 25,000 K and its diameter is 11,700 km. This star of 1.034 solar masses is actually smaller than the Earth!

10.2. Pauli's exclusion principle states that no two electrons in an atom can have the same four quantum numbers that specify its energy and angular momentum states.

10.3. If the white dwarf in a binary accumulates sufficient mass to bring it above the Chandrasekhar limit, the dwarf would begin to collapse under its own gravity. Current models indicate instead that before that can happen, nuclear fusion ignites in its interior. The dwarf immediately disintegrates in a catastrophic explosion. The outcome is a Type Ia supernova. Type Ia supernovas have the interesting property that they all have the same maximum brightness. That makes them suitable as "standard candles" that can be used to determine cosmic distances.

10.4. The Sun's mass equals the mass of 10^{57} neutrons; each neutron has a diameter of about 10^{-13} cm. So a sphere of diameter 10 km would contain all of them if they were tightly packed.

10.5. For this discovery, Hewish shared the 1974 Nobel Prize in Physics with Martin Ryle. Bell was thanked but did not share the prize. The injustice of the award stirred a vigorous controversy. Bell was awarded many prizes later but none with the prestige of the Nobel.

10.6. Gamma ray bursters are thought to be neutron stars in the act of collapsing to a black hole. They appear at random positions in the sky about once a day.

10.7. The network now includes the Italian Satellite for Astronomy X (SAX); the U.S. Rossi X-ray Timing Explorer (RXTE); the U.S. Chandra X-ray Observatory (CXO); and the European Space Agency Multi-Mirror Observatory (XMM-Newton).

10.8. A distant galaxy appears to rotate westward with the stars, due to the Earth's rotation. But in its orbital motion, the Moon moves eastward 13.2 degrees per day against the background of the stars. That means that a distant galaxy in the Moon's path will eventually be eclipsed.

10.9. A source of light that varies in, say, T seconds cannot be much larger than the distance $D = c \times T$, where c is the speed of light. Otherwise, light from more distant parts of the source would arrive much later than the light from the nearer parts and the overall time for the variation would be longer than T.

10.10. The existence of a black hole at the center of the Milky Way was confirmed in 1998, when A. M. Ghez and others observed the radial velocities of stars within 0.1 arcseconds of Sagittarius A, the radio source at the center of the Milky Way. From the speed and orbits of the stars, they could derive the hole's mass. Two years earlier Reinhard Genzel and an international team obtained an independent estimate. These teams agreed that the mass of the hole equals 2.6 million solar masses.

10.11. When a fast electron collides with a low-energy photon, it can deliver part of its kinetic energy to the photon. The photon can become a hard x-ray through the transaction. This is the inverse Compton effect. In the Compton effect, an energetic photon boosts the energy of an electron, possibly to relativistic levels.

11. THE GALAXIES

11.1. A parsec is the distance at which the diameter of the Earth's orbit around the Sun spans an angle of 1 second of arc.

11.2. An electron in a hydrogen atom has a spin due to its rotation and also a spin due to its revolution about the proton. If the two spin directions are in opposite directions, the atom has slightly less energy than if they are aligned. Whenever the electron jumps spontaneously to its lower energy state, it emits a photon at a wavelength of 21 cm. This is the telltale spectral line that van de Hulst discovered.

11.3. Actually, the orbits of several satellites precess about their parent planets. For example, the Moon's orbit precesses about the Earth due to the gravity of the Sun, completing a revolution in 18.6 years.

11.4. Light is an electromagnetic wave, in which an electric field vector and a magnetic field vector oscillate at right angles. In an *incoherent* source, like a light bulb, the vectors rotate rapidly and randomly about the line of sight; the light is *unpolarized*. In some sources, like the glare from the ocean, the light is *partially polarized*. Starlight is also partially polarized. If we look at it through a sheet of Polaroid, it selects only the instants at which the electric vector aligns with the axis of the sheet and we see the *plane-polarized* component of the light. Some sources, like lasers, emit coherent light, which is completely polarized.

11.5. Davis and Greenstein pointed out that random collisions with gas atoms will cause a micron-sized dust grain to spin at speeds of 10^9 rpm. If the grain is not aligned with the magnetic field, it sees a fluctuating electric field that dissipates part of its rotational energy. The result is a torque that aligns the axis of rotation with the field.

11.6. One also needs the distance to each pulsar. These distances are relatively easy to acquire from trigonometry.

12. SOMETHING FROM NOTHING

12.1. The redshift, z, of a source is defined as the difference between the observed and initial wavelengths of the spectral line, divided by the initial wavelength. So, for example, if an identified line was emitted at a wavelength of 120 nm by a source and observed at 600 nm, the redshift z would be equal to 4.0. The age and distance of the source must be calculated from a model of the expansion of the universe, which may contain uncertain parameters. Therefore, researchers commonly refer tentatively to a source's age or distance by quoting its redshift.

12.2. In this academic example, the force on the selected proton due to the temperature gradient is the result of more energetic collisions from the right side than from the left side. Particles to the right of the selected proton are hotter (more energetic) and exert more force on the proton, through collisions, than those to the left. Hence the proton would feel a net force to the left. The same is true for the selected electron.

INDEX

Credits

Fig. 2.1: Courtesy of Dave Ewoldt, Okarche, Oklahoma, www.okweatherwatch.com

Fig. 2.4: Courtesy of G. A. Glatzmaier and P. H. Roberts, *Contemporary Physics* 38, 269–88, 1997

Fig. 3.1: Courtesy of TRACE satellite

Fig. 3.2: Courtesy of David Hathaway, NASA Marshall Space Flight Center

Fig. 3.3: Courtesy of M. Stix, *Astronomy and Astrophysics* 47, 243, 1976

Fig. 3.4: Courtesy of Friedrich Woeger, NSO/AURA/NSF

Fig. 4.1: Courtesy of TRACE team

Fig. 4.3: Courtesy of NASA and the LASCO team

Fig. 4.4: Courtesy of P. A. Sturrock; published in *Structure and Development of Solar Active Regions,* ed. Karl Otto Kiepenheuer, International Astronomical Union (Dordrecht: D. Reidel), 471

Fig. 4.5: Courtesy of T. Amari et al., *Astrophysical Journal* 529, L49, 2000

Fig. 5.1: Courtesy of NASA

Fig. 5.2: Courtesy of G. Pneuman and R. Kopp, *Solar Physics* 18, 258, 1971

Fig. 5.3: Courtesy of J. R. Jokipii and Barry Thomas, *Astrophysical Journal* 243, 115, 1981

Fig. 5.4: Courtesy of NASA

Fig. 6.1: Author; drawn from James Dungey, *Planetary Space Science* 10 (1963): 233

Fig. 6.2: Courtesy of P. H. Reiff, Rice University

Fig. 6.3: Courtesy of Stanley W. H. Cowley, *Earth in Space* 8, no. 7, 6, 1996

Fig. 6.4: Courtesy of C. Xiao et al., *Nature Physics* 2, 478–83, 2006

Fig. 7.1: Courtesy of J. E. P. Connerney et al., *Geophysics Research Letters* 28, no. 21, 4015–18, 2001

Figs. 7.2–7.3: Courtesy of F. Bagenal, *Annual Review of Earth Planetary Science* 20, 289, 1992

Fig. 8.1: Courtesy of NASA

Fig. 8.2: A.-K. Jappsen et al., *Astronomy and Astrophysics* 435, no. 2, 611–23, fig. 3, 2005

Fig. 8.3: Courtesy of Frank Shu, *Astrophysical Journal* 429, 721, 1994

Fig. 9.1: Courtesy of S. Vogt and G. D. Penrod, *Publications of the Astronomical Society of the Pacific* 95, 565, 1983

Fig. 9.2: Courtesy of O. Kochukhov, *ASP Conference Series* 305, 90, fig. 5, 2003

Fig. 9.3: Courtesy of O. Kochukhov, in *Proceedings IAU Symposium* 224, 436, fig. 3, 2004

Fig. 10.1: Courtesy of P. Goldreich and W. Julian, *Astrophysical Journal* 157, 869, 1969

Fig. 10.2: Courtesy of A. Spitkovsky, *Astrophysical Journal* 658, L51, 2006

Fig. 10.3: Courtesy of R. Blandford and R. Znajek, *Monthly Notices of the Royal Astronomical Society* 179, 433, 1977

Fig. 10.4: Courtesy of R. Perley et al., *Astrophysical Journal* 285, L35, 1984

Fig. 11.1: Courtesy of R. Beck, *EAS Publications Series* 23, 19–36, 2007

Fig. 11.2: Author; after A. J. Kalnajs, *Proceedings of the Astronomical Society of Australia*, 2, 174, 1973

Fig. 11.4: Courtesy of J. C. Brown et al., *Astrophysical Journal* 663, 258, 2007

Fig. 11.5: Courtesy of D. Moss et al., *Astronomy and Astrophysics* 380, 55, 2001

Fig. 12.1: Author, after L. M. Widrow, *Reviews of Modern Physics* 74, 775, 2002